Writing and Designing Manuals

SECOND EDITION

Operator Manuals

Service Manuals

Manuals for International Markets

Gretchen Holstein Schoff
Patricia A. Robinson

 LEWIS PUBLISHERS

Library of Congress Cataloging-in-Publication Data

Schoff, Gretchen H., 1931–
  Writing and designing manuals : operator manuals, service man-
uals, manuals for international markets / Gretchen Holstein Schoff,
Patricia A. Robinson. -- 2nd ed.
    p. cm.
    Rev. ed. of: Writing & designing operator manuals. c1984.
    Includes bibliographical references and index.
    ISBN 0-87371-421-0
    1. Technical writing.   I. Robinson, Patricia A., 1948-   .   II.
  Schoff, Gretchen H., 1931–   Writing & designing operator man-
  uals.   III. Title.
T11.S376 1991
808′.0666--dc20   91-8525
                                                                          CIP

ISBN 0-87371-421-0

LEWIS PUBLISHERS, INC.
121 South Main Street, P.O. Drawer 519, Chelsea, Michigan 48118

PRINTED IN THE UNITED STATES OF AMERICA

Preface

"If you want to find out the best way to sew on pockets, don't ask the president of the shirt company. Ask the person working on the line, someone who has sewn on 10,000 pockets."

We wrote the first edition of this book with that advice in mind and have continued that philosophy with this new and expanded second edition. The book is a practical, "how-to" book drawn from our experience teaching and consulting with technical writers who produce the operator and service manuals for a wide array of consumer and industrial products. Since that first edition, the technical publication enterprise within industry, business, and government has grown steadily in complexity, sophistication, and importance. The book reflects these new trends and incorporates many new materials on desktop publishing, techniques for visuals, safety and product liability, and production of translated manuals for international markets. We have included many new sample manual pages, drawing from our growing collection of manuals provided to us by product manufacturers. Not only have manual writers shared their work with us—they have educated us on the realities and challenges of their profession. So when we give suggestions, they are usually a compendium of the shared wisdom of writers who have tried the techniques on the job. They are the people who have sewn on the "10,000 pockets" producing the thousands of manuals and instruction books that accompany products. We owe them thanks for what they have taught us.

We have organized the book along the same lines as the major steps taken in writing the manual: planning; analyzing the user; choosing organizational and writing strategies; coordinating format, references, and mechanics; and creating visuals. We also include chapters on the following special topics: safety messages, service manuals, manuals for international markets, and manual production. Writers make many choices as they put together a manual. We describe the decisions that must be made and show which techniques work best. Careful control of the manual elements produces a document that is clear, technically correct, and readable by the widest possible range of users. Throughout our book—but especially in Chapter 1, Planning, and Chapter 9, Managing and Supervising Manual Production—we discuss techniques for organizing office structures and writing teams to make the manual production process run as smoothly as possible.

We have tried to keep the language of this book clear, direct, and easily understandable. The principles it discusses can be applied in writing manuals for a diversity of products (household, automotive, chemical, industrial, mechanical, recreational, biomedical). Throughout the book, you will find many examples and illustrations. For instance, the chapter on analysis of user needs provides a series of questions that can be used to make such an analysis. The chapter on visuals provides samples of effective photos and drawings, plus advice on how to produce them. The chapter on safety warnings shows how to design them and how to avoid ambiguous wording.

If, as a writer on the job, you are faced with a deadline, and you are writing copy, choosing paper and type size, hiring a printer, laying out pages, producing drawings and photos (or contracting for someone to do them)—this book will help you produce a manual that helps to protect both the user and the manufacturer of your product.

Gretchen Holstein Schoff
Patricia A. Robinson

The Authors

Gretchen Schoff is a Professor in the Department of Engineering Professional Development, the Department of Integrated Liberal Studies, and the Institute for Environmental Studies at the University of Wisconsin-Madison. Professor Schoff teaches writing and interdisciplinary courses on campus, has consulted for industry and government in the U.S. and Canada, and has taught continuing education seminars for industry. Professor Schoff has also served as Editor for the Center for Engineering Research at the University of Hawaii. At UW-Madison, she is past Chairman of the Institute of Environmental Studies and is currently Associate Chairman of the Department of Integrated Liberal Studies, where she also directs its writing program. In 1980 she received the Chancellor's Distinguished Teaching Award.

Patricia Robinson is Assistant Professor in the Department of Engineering Professional Development at the University of Wisconsin-Madison where she teaches campus courses and directs outreach programs in technical communication. She is Manager of Campus Courses for the department and past Director of the Technical Communication Certificate program. She also teaches seminars for technical writing practitioners and consults with industry.

Professor Robinson has taught writing at the college level for 17 years and since 1978 has concentrated on technical writing. She is the author of *Fundamentals of Technical Writing* and is co-author of *Effective Writing Strategies for Engineers and Scientists*. She is particularly interested in the role of instructions and warnings in safety and liability prevention.

Contents

List of Figures

Acknowledgments

Many individuals have contributed to this book by sharing their expertise and their materials or by allowing us to observe and work with their writers. We owe a special thank you to the following people: John Conrads, formerly of Deere and Company; Dale Fierke, Tetra Pak; Charlie Freeman, Hewlett-Packard; John Gormley, Westinghouse Electric Corporation; David J. Howard, Clark Components International; Albert O. Hughes, FMC Corporation; Bill Lichty, Scotsman; T.W. Loetzbeier, Mack Trucks; David S. Maslowski, The Kartridg Pak Company; Jimmie Moeller, Gehl Company; Richard Moll, Department of Engineering Professional Development, University of Wisconsin–Madison; Fred Rode, Outboard Marine Corporation; Pete Shelley, General Electric Medical Systems; Delmar Swann, E. I. DuPont De Nemours and Company; Stan Sweetack, Pierce Manufacturing Inc.; John Thauberger, Prairie Agricultural Machinery Institute; Ivan Thue, Prairie Implement Manufacturers' Association; George Winkleman, Delta International Machinery Group.

The materials for the book were collected in a number of ways. We have conducted in-house technical writing seminars for industries, visited service publications operations, lectured and taught in national and regional conferences and workshops devoted to product safety and technical writing, and served as private consultants. Individuals and organizations who have worked with us and/or whose materials provided the subjects and examples for this book include Acme Burgess, Inc.; American Optical Company; Atwood Mobile Products; Butler Manufacturing Company; Chrysler Motors Corporation; Clark Components International; Construction Industry Manufacturers' Association; CooperVision Surgical Systems; Deere and Company; Delta International Machinery Corporation; Devilbiss Company; Doboy Packaging Machinery; E. I. DuPont De Nemours and Company; Farm Implement and Equipment Institute; FMC Corporation; Ford Motor Company, Ford Tractor Division; Ford New Holland; Fulton Manufacturing Corporation; Gardner Denver; Gehl Company; General Electric Company and General Electric Medical Systems; General Motors Company; Gerber Products Company; Harley Davidson Motor Company; Hazelton-Raltech Incorporated; Hewlett-Packard; Honeywell Inc.; Huffy Corporation; Ingersoll-Rand Company; International Business Machines; International Harvester Company; J. I. Case Company; John Muir Publications; Joy Manufacturing Co.; Kartridg Pak Company; Kohler Corporation; Krones Inc.; Mack Trucks-Mack International; Madison-Kipp Corporation; Martin Engineering; Norden Laboratories; Ohio Medical Products; Outboard Marine Corporation; Pierce Manufacturing Company; Prairie Agricultural Machinery Institute; Prairie Implement Manufacturers' Association (and affiliated companies of Canada); Rosemount Inc.; Scotsman; Siemens Medical System; Silver-Reed America, Inc.; Teresa Sprecher; Rosemary Stachel; Sunstrand Aviation; TRW Ross Gear Division; Taylor Instrument Company; Technicare; Thern Inc.; Versatile Corporation; Volkswagen of America; Wabco Construction and Mining Equipment; Westinghouse Electric Corporation; and Yahama Motor Company.

1

Planning

Overview

Manual writers from many businesses have been our guides in creating this book. They keep asking questions that lead us to look for answers, and they have generously shared with us their favorite tricks of the trade. We have also found that whether writers are preparing manuals for copiers or paint sprayers, for cash registers or trailer hitches, they encounter similar problems. Writers often see those problems as the direct result of poor planning before actual manual production starts.

This chapter treats the planning process. Whether manuals are produced by a single writer or by a team, careful planning does much to set the stage for effective manual production. We begin by looking at the planning involved in fitting manual production into company structures. Then we look at the planning necessary for technical writers who work singly or in teams and at preliminary planning for the manual itself. Because writers need information and time, we give suggestions on how to fulfill these basic needs. Many of these suggested techniques can be used not only by writers, but by publications managers and supervisors as well (see Chapter 9, Managing and Supervising Manual Production, for fuller treatment of supervisory techniques).

Manuals: Their Function and "Fit" within a Company

Operator manuals give verbal and visual instructions for the use of thousands of products, ranging from toasters and tractors to cameras and cash registers. Almost every product except the simplest comes with such instructions. These instructions are known as operator manuals, owner manuals, user manuals, or simply instructions. They are "how to" books for owners and operators of the products.

Technical publications, including manuals, are a critical part of company organization, but before you do planning for manual production, you should ask a key question: "How do technical publications fit into your company organization?" Asked this question, technical writers we have talked to are sometimes at a loss to describe the "fit" of their department within the company. This happens especially where manual production has grown rapidly over the years, with haphazard, unplanned responses to crises and needs, but no real plan for technical publications. In talking with writers and managers of writers, we have found these to be the most common structures:

- A stand-alone technical publications department responsible for all company products
- Technical publications subsumed under another division, such as Engineering, Safety, Marketing, or Service
- Separate technical publication units for each product (often located at widely separated geographic locations or within separate divisions)
- Single or small groups of writers who produce all the documentation, but are unattached to any unit except by budget. (They are literally "one-man bands.")

The Importance of Manuals as Publications

At first glance, operator manuals and service manuals may seem to involve little more than instructions for use and care of the product. In reality, manuals do much more.

Instructions

Most operator manuals contain instructions for assembly, operation, maintenance, and storage of products. Many manuals also contain sections on trouble-shooting, service, and repair. Very complex products, such as electronics, heavy industrial machinery, or biomedical devices, usually have separate manuals for service and repair. New products, for which no prototype manuals exist, and complex products, in which many subsystems interact, present especially difficult problems for writers.

Such manuals demand that the writers have a grasp of an astonishing diversity of mechanisms, processes, and procedures. Therefore, writers must continually be asking themselves, "How can I best describe or show how this works?"

Product Liability Document

The consumer protection movement and the legal climate surrounding product liability law are of serious concern to manufacturers of any product. Of special current interest are the safety warnings found throughout operator manuals and the safety labels actually affixed to the product. One important component of product liability law is the manufacturer's "duty to warn." Manuals must warn product users against such hazards as electrical fields, sharp blades, moving parts, shattering glass, chemicals, toxic substances, and flammable and explosive materials. The operator manual and its warnings frequently become key documents in product liability suits. Therefore, if the manual is well designed and worded, it may help protect the manufacturer against charges of failure to give adequate warning.

An Advertisement for the Product and a Part of the Company Image

The operator manual sends a message to the buyer of your product. A poorly designed, confusing, or unreadable manual may cast doubt on the quality of the product itself or convince a user not to buy from a certain company in the future, even if the products of that company are good. Think, for instance, of a buyer who tries out his new camera for the first time and finds the manual so hard to use that he can't put the film in the camera without returning to the photo shop. His teeth are set on edge even before he has had a chance to try the product. In contrast, top-quality manuals are read, used, and saved by the owner—and often envied and copied by competitors.

Establishing Links

There are many permutations of company organizational patterns, and each permutation has advantages as well as special problems. If you have not looked at the relationship between technical publications, as a whole, and the rest of your company structure, you should. When you do this, you will not only be able to plan more effectively, but you will also understand problems better, as well as see opportunities for changes and improvements. You will need to consider the following elements in your company's overall structure as well as within the technical publications unit itself.

Elements to Consider in Assessing Operations of Technical Publications

- Organization—work flow relationship of company units to each other
- Management—lines of authority, decision making
- Personnel—selection, numbers, compensation
- Scheduling—work flow and deadlines
- Budgeting
- Networking—contacts, information sharing, duty sharing
- Publications—manuals, service, training, brochures, sales, specification sheets, parts catalogs

These elements of planning give evidence of professionalism. A technical publication is professional when it applies the same kinds of coordination and accountability used elsewhere in the company. Quite often, the Technical Publications Department lies at the crossroads of several company operations, the place where sales literature, service and operator manuals, parts lists, training manuals, specifications, graphics, video, film, and computer often come together. Positioned at the crossroads, as technical writers frequently are, they have unique opportunities to forge links with many other major units in the company.

Writers at the Crossroads

On a good day, manual writers might describe their jobs as "being at the crossroads." On a bad day, the same writers feel more as if they are working at a dangerous intersection, with traffic coming in all directions. Writing manuals is tough, messy, and chaotic. That piece of wisdom is probably the most important lesson taught us by writers who do the job. So when we talk here about planning and organization of manual writing, we do so remembering what writers have told us about the realities of their work. The planning techniques we suggest have helped writers bring some order into a job that has considerable chaos built in—it goes with the territory.

Mission Impossible: The Role of the Technical Writer

We are convinced that the technical writer whose principal task is producing manuals has one of the most difficult jobs around. It is difficult because the writer is responsible for a crucial task—producing a manual that is the company's main connection with the customer (outside of the product itself)—yet he or she seldom has the full authority needed to do that task well. Indeed, many of the major decisions affecting the production of the manual, including both content and schedule, often are made by persons in other areas of the company. Deadlines may be set by marketing to coincide with a new model period without regard to the complexity of the writing task; information needed to meet those deadlines may be held up in engineering because of last-minute design changes. Yet the technical writer is expected to produce usable, accurate manuals, on time and within budget.

In a later chapter, we examine what a manager should look for in hiring technical writers. What is clear from the outset is that neither the "technical" nor the "writer" part of the job title is as important an attribute as a candidate's "people skills." The technical writer depends on good relationships with people in other departments—engineering, marketing, manufacturing, and service, among others. The people skills are so important because the technical writer is seldom the subject-matter expert—instead, the writer relies on others for the information that provides the substance of the manual.

Equally important, we believe, is the ability to live with a little chaos. Over a dozen years we have met and talked to hundreds of technical writers. We have not yet found a single one who thought his or her job was straightforward and easy. Instead, they describe it as a juggling act in which they are always short of two key elements: time and information. In the real world, writing manuals will never be accomplished in the way your English teacher told you to write a term paper.

The Writing Process
(as taught by the English teacher)

1. Make basic document design decisions

   Product coverage
   Format
   Schedule

2. Gather information

   About the product
   About the users

3. Prepare outline and list of visuals
4. Write draft and prepare visuals

5. Edit and get approvals
6. Typeset, paste up, and print the manual

Instead, it will always look more like the following:

The Writing Process
(as it really happens)

1. Receive assignment with nearly reasonable deadline
2. Begin making basic decisions
3. Deadline moved up 2 weeks
4. Try to get information from engineers on product; receive specification sheet with illegible handwritten changes
5. Try to get product; receive outdated model with parts missing
6. Deadline moved up 2 weeks
7. Start to write anyway; receive current prototype; celebrate
8. Overhear lunchroom conversation about radical design changes in product; scrap draft
9. Deadline moved up 2 weeks

The Solo Writer

Small companies often assign one person to do the manuals for products. If you are that solo writer, you will soon find that writing is only one part of the job. You may also have to do photography, plan the art work, choose paper stock, edit, type, and desktop publish.

Advantages

As a solo writer, you have many opportunities to be creative. Because the majority of decisions will fall to you alone, you can approach the manual production job with your own vision of how the final manual will look, and you can make certain decisions without having to clear each step of the production with someone else. We have met a number of solo writers who say that the autonomy they enjoy more than compensates for their many responsibilities. They also value the variety of tasks involved and enjoy the different kinds of people they work with. Most of all, they like having control over the project from start to finish.

Disadvantages

If you are a solo writer, your work will be the single bridge between the technical data about your product and the manual that reaches users. You will have to gather the information and create the schedule yourself. Manual writers who work solo often feel rushed, isolated, and pressured by their many responsibilities. They sometimes feel that other personnel, those on whom they must rely for information, have scant understanding of what it takes to put a manual together.

Making Solo Writing Easier

Much depends on your ability to handle solo writing in a professional manner, but you have the disadvantage of not having a writing team to lean on, either for moral support or advice. When you work alone, one of the most logical and understandable ways to get others to understand your work is to "keep books" on what you do. For instance, you might begin by simply keeping a record of how many hours it usually takes you to create a manual page for a new product or how much time it takes to produce a computer-generated graphic. Even relatively uncomplicated bookkeeping can be a big help if you have some facts and figures to show that deadlines are unreasonable, costs are too high, or one person can't do the job alone.

Your needs as a solo writer are much the same as the needs of team writers. You need access to information and time to do the job. As you read the rest of the chapter on team writing and on information and time, you will find many suggestions that you can adapt to the solo-writing setting.

Consider, especially, ways in which you can perform the same functions as a team leader performs in team writing. For example, you can do your own advance planning by:

- Arranging your own schedule of meetings with key personnel to collect information
- Asking for help from informal support teams or individuals (for work such as typing, drawing, planning safety messages, taking photographs)
- Developing a thorough outline
- Laying out steps in manual production
- Setting up a style and format handbook or a set of guidelines so that your own writing procedures become standardized and easier to repeat from manual to manual

Team Writing

Manual writing is often done by a team of writers, especially if the product is complex. Such division of labor makes sense for a number of reasons: preparation times can be shortened, writers can develop special expertise with certain manual segments, and teams can include personnel from other company units (e.g., technical, research, product safety). The team-written manual also poses problems, particularly those of conceptual unity, team coordination, and uniformity of quality. Here are some of the pros and cons of the team-written manual and some suggestions for making team writing efforts smoother.

Advantages

"Many hands make light work" is a familiar saying. Dividing the manual writing according to systems or processes inherent in the product or according to special areas of writer expertise allows you to make the best use of writer talent and to get the job done more quickly and accurately. For example, the writer whose specialty is filtration systems, calibration, or electrical systems will find it easier to write about that area than the writer who has to keep many different kinds of processes or procedures in mind. Situations also arise in which a machine or product has used standard mechanical or chemical processes and is then suddenly altered by new technology or by the addition of an electronic component. In such cases, the best use of talent may be to ask the technician or engineer-designer who created the new component to write the segment describing its function. The most frequent kinds of product alterations in the last decade have been those involving computers, numerical control, or robotics (e.g., devices for welding, spraying, assembly procedures, and quality control).

Disadvantages

"The camel is an animal designed by a committee" is another saying. Too often, the team-written manual has camel-like lumps and bumps. Such manuals move by fits and starts from one segment to another. They sometimes have ill-matched writing styles and formats. Users find these manuals very hard to use because of their redundancy, lack of cross-referencing, and chaotic organization. In brief, the chief difficulty with the team-written manual is the coordination of several writers' work into a smooth manual that looks as if one person had written it.

Coordinating the Team Effort

The team-written manual is a reflection of company structures and procedures, as well as management styles and individual personalities. Coordination of team writing should make the best use of available time, talent, and money. "Who decides what, and when?"

Team-written manuals often have a team leader, a manager, or a service publications editor who has the final responsibility for and a unified concept of what the finished manual will look like. That unified concept may be the joint creation of the writing team; however, once the conceptual framework is established, leaders are often responsible for the scheduling, assignment of manual segments, creation of clear instructions for what each manual segment is to include, and final coordination and editing of the completed manual.

Team leaders should have strong writing and editorial skills because they will have the job of making language, style, and format internally consistent. A good team leader will make use of instructions, writer guidelines, writer checklists—any procedure that helps writers know what is expected of them and when. In the last section of this chapter, we have provided some samples of checklists, work schedules, and information-gathering techniques. These may be adapted to match your company's structure and procedures.

What the Writer Needs

Writers, whether in small or large companies, have some very basic needs. To do a good job, they must have access to information and adequate time.

Information

Believe it or not, we have met many writers whose chief frustration was a lack of information about the product. They may ask to see the product and be refused. They may ask for scheduled time to review the product with designers, technicians, engineers, or safety personnel and be told that there is no time. They may ask for a working model, a prototype, or at least a photo and get a flat "no" for an answer—or they may be housed in an office miles away from where the product is produced. Admittedly, many people may be clamoring for a prototype of a new product. Marketing wants it, engineering is working on it, and product safety needs it. When the pressure is on and deadlines must be met, writers often get short shrift. However, if the manual is to perform its function, writers must have information and management must provide the procedures to help them obtain it. Information gathering is an important first step in planning the manual.

Time

Our comments on time as a basic writer's tool are directed especially to managers. Deadlines are the name of the game in most industries. More errors and slapdash jobs can be explained by time pressures than by incompetence. Managers often need to be reminded that writing takes time. Writers themselves usually do not have to be convinced. They know that writing takes more time than anyone would ever guess, though they sometimes underestimate how much. We have surveyed writers in our seminars who estimate that even an average one- or two-page business letter or memo may take several hours to compose.

Once writers and/or supervisors have created effective information collection systems and have gone through the manual production process at least once, subsequent manual production proceeds more quickly. However, totally new products need especially generous lead time for creating the manual, since some of the vital information may not be available until the last minute, when the prototype is completed and tested.

Information: How to Collect It

People tend to think of writing as a solitary occupation—the writer alone with resource material and a word processor hammering out paragraphs of golden prose. The actual writing may be solitary, but for technical writers, the resource material is almost always other people. A good manual writer spends much of the workday out and about, talking to people—engineers, technicians, service experts—because they have the information. However, these people are all busy with their own jobs; how can you get them to take time to help you?

Tactics

In a very small company, the writer may need only to lean across the desk and ask a co-worker for information. In very large companies, information collection is more complicated. Successful writers we have talked to mention three key tactics for coping with these obstacles. (Note that we did not say "solving these problems"—they will never be totally solved. Writing manuals is like shooting at a moving target: products constantly change to incorporate innovation or meet changing market needs; the manuals that accompany those products must also change.) These three tactics, however, will help reduce the chaos:

1. *Make yourself part of the product development team.*

   The earlier you can be involved in the product development effort, the earlier you can start writing the manual and the more timely information you will have. If your company does not already have the technical publications department as a part of product development, volunteer. As one technical writer put it, "I just started going to meetings to which I wasn't invited. After a while, they expected me to be there!"

2. *Cultivate contacts in key areas*

   Develop relationships with one or two people in engineering, service, and so on. If you cultivate a good working relationship with someone, that person will be more likely to pick up the phone and call you to tell you about a design change rather than to wait passively until you come to him. Developing these contacts may mean trading favors: as the writing expert, you may be asked to edit letters or look at documents that are outside your job description. When you can, be friendly and helpful—you'll be in a better position to ask for help when you need it.

3. *Develop the manual as a series of modules.*

   One of the writing techniques we have found to be helpful is "modular" writing, in which the manual is conceived of as a series of nearly self-contained segments. Some writers go so far as to limit the size of these segments to no more than a two-page spread, since that is all that the reader can see at one time. However you define a module, the principle is the same: break the writing task down into bite-size units. This way you can work on the different sections independent of one another. Even if you don't have all the information for Section A, you can still work on Section B. In addition, the modular writing technique makes it easier to revise manuals (you can substitute new modules for those affected by model changes and leave others untouched) and to use relevant modules in more than one manual.

These suggestions will help to keep your job manageable. Adapt your information-gathering systems to the realities of your company and be on the lookout for ways to help the right people talk to each other. Listed below are some of the successful techniques used by industries to improve information collection.

Information-Gathering Aids

- Product development meetings (Product development meetings are one of the richest sources of information for manual writers. When manual writers participate, from earliest phase to final product, in these meetings, they have a steady flow of information that gets plugged into the manual.)
- A product safety committee that includes writer representatives when key safety features and messages are being discussed
- Placement of writers' offices near production, research, and test facilities (This assures that the writers are not working in a vacuum. One look at a product is worth 20 phone calls.)
- An orderly file system that allows the writer to reuse materials and modules prepared for other manuals, especially if the manual describes a new product with only minor design changes or a slight model change
- A product history that alerts writers to places where bad manual writing may have caused operator problems, accidents, or death
- A scheduled walk around the product or prototype well in advance of the manual deadline
- Writer guidelines and style handbooks prepared in-house by editors
- A writer's checklist that includes these items:

  Specifications and dimensions
  Brief description of product function and use
  Important safety features and hazards
  New design features unfamiliar to the writer

  (This information is provided by engineering, marketing, product safety, and/or testing — whatever group bears responsibility or has the information.)

All of the suggestions listed above are subsumed under a single prerequisite: *writers must have access to information.*

Time: How to Schedule It

It would be wonderful if, when a company planned to develop a new product, someone came to the Technical Publications Department and said, "How long will you need to develop the appropriate documentation for this product?" Instead, Marketing sets the shipping date in consultation with Engineering and Manufacturing . . . and Technical Publications is expected to have the manual ready to ship with the product. However, writing takes time; how can you develop the manual in parallel with the product?

Manual production becomes much easier if, at the planning stage, you can determine approximately how much time you can devote to each part of the production process. A product is rarely allowed to leave the production facility until the accompanying manuals or instructions are complete and ready for distribution or packaging with the product. The controlling date, then, is the deadline when the product is scheduled to be shipped or sold. Establish this final deadline and create a work-flow schedule that allots time for the following:

Phase 1: Initial Planning

- Clarifying the manuals' functions
- User analysis
- Outline and/or storyboard development
- Writer assignments
- Information collecting
- Writer checklists and guidelines

Phase 2: Preparing the Manual

- Writing and layout
- Creation of visuals
- Reviewing
- Editing
- Revision
- Preparation of final copy
- Printing

Tasks in Manual Production

The following sections describe the activities involved in each of the tasks of manual production. The sequencing and overlapping of the activities listed above vary enormously from company to company. Here is a thumbnail sketch of what is typically involved in each of the tasks.

Phase 1: Initial Planning

Clarifying Manuals' Functions

Manuals are seldom "stand alone" documents. They are usually part of an array or a family of supporting documentation for the product. Think about how the manual fits into the documentation family. If you think about the manual as part of the cluster of documentation, you are more likely to have complete coverage of all the essential areas as well as to greatly reduce later headaches arising from overlap and repetition.

User Analysis

Steps for user analysis are described in Chapter 2, Analyzing the Manual User. Analyze the user before writing the manual—this will allow you to decide on appropriate format and language levels for the manual. User questions may be employed to develop the outline and to determine what major sections the manual should contain.

Develop Outlines or Storyboards

Develop outlines or storyboards for major sections and/or chapters of the manual. Employ user questions to help you establish major sections of the manual (see Chapter 2). These outlines should be detailed enough to help you refine cross-referencing and to prevent redundancy and over- lap. Good outlines let you see major blocks and chunks of information before you commit yourself to the actual writing. These chunks or mod- ules are much easier to move around and refine at the outline stage, than they are after you start writing actual text.

Writer Assignments

These may be for a complete manual or for segments of a manual. Decide whether to use solo writers or writing teams and assign responsi- bilities for editing the manual and for making final judgments on bind- ing, paper stock, and page size.

Information Collecting

Sources of information may be product designers, engineers, and person- nel in marketing, sales, research and development, product safety, and production. Other sources may include product histories, files, photos, test data, and reusable materials from other manuals. Information gath- ering goes on from start to finish in manual production.

Writer Checklists and Guidelines

These may be prepared by the writer(s) and/or supervisors. Guidelines include lists of specifications and dimensions, special safety hazards, and new design features of the product. These may also include standard in–house instructions for format, standardized glossaries of terms for parts and procedures, and the types and sizes of photos and drawings that may be used.

Phase 2: Preparing the Manual

The rest of this book deals in considerable detail with such activities as writing and layout, visuals, and the editing and revision that make up final design of the manual. Preparation of final copy and printing choices depend, of course, on whether you are publishing in-house or by outside contractor.

Phases of Manual Production

Manual writing is seldom a series of smooth steps. Activities overlap and often go on simultaneously. Those activities most often squeezed out or omitted are clarification of manual function, outline development, and user analysis. Time spent on these is not time wasted but time gained because it sets up the work place and establishes a clear picture of manual function.

Phases of activity will vary in their sequencing, depending on the size of the company and its organizational structure. For example, some companies have not one but several review or editing steps that may come early or late in the work-flow schedule. Further, the creation of visuals, the planning of format, and the writing of verbal text usually occur simultaneously. Finally, the best of schedules will develop glitches — missing data, late photos, key people sick or out of town, and/or last-minute design changes.

In short, you need to sequence and overlap the steps we suggest in a way that makes sense for your company, but do not neglect or omit Phase 1 (see Chapters 3, 4, and 5 on writing strategies, formatting, and visuals).

Scheduling Responsibility

In large companies, manual scheduling is usually handled by team leaders or publications editors. If you are working as a solo writer, you will have to do much of the scheduling yourself. Schedule yourself extra time when you have to collect information from other people or when you have arranged for outside help (e.g., with typing, printing, or drawing).

Summary

Planning is the first step to successful manual production. It begins with intelligent allocation of time, money, and personnel and with organization of the work place. Writers need information and time. Do all you can to fulfill these basic needs and regard information collection, user analysis (Chapter 2), scheduling, and outline development as the essential groundwork for manual writing.

2

Analyzing the Manual User

Overview

User analysis should be included as one of the tasks in planning and producing manuals. Writers should have a clear picture of manual users and their needs. What the user needs to know should guide manual writers in choices of language level, reading level, safety warnings, execution of visuals and graphics, and arrangement of manual segments—all the elements discussed in subsequent chapters of this book.

This chapter gives guidelines for user analysis and user feedback. We describe the spectrum of manual users, from sophisticated to naive, and show the differences between the professional and the general public user.

We provide a list of products most likely to be bought by the general public and some questions to use as guides in analyzing your user. Then we provide a list of questions typically asked by users. We show how these questions can help generate manual outlines and decide on major sections of the manual. The last segments of the chapter suggest techniques for collecting user feedback. We show how to simulate the person-on-the-street user as a way of avoiding "shop blindness"—the inability to see your products as a first-time user might—and we show how to make use of feedback in revising and updating manuals.

At the end of the chapter, you will find a checklist to help analyze your user.

The writer has manipulated several language elements, adjusting the passage with specific users in mind.

Quite often, the same product will have two manuals, written at different language levels, or it may have separate manual segments, some written for professionals and some written for general public users. These two passages could both occur in the same manual — the first in a segment directed to surgeons or trained technical personnel and the second in a segment directed to equipment assistants and nontechnical personnel. (The service manual for professionals relies on the user's in-depth knowledge of the product, but also demands that the manual be put together somewhat differently than a manual for the general public. We deal with this special kind of manual in Chapter 7, Service Manuals.)

Comment. Both passages describe the same mechanism. Passage A assumes the reader knows the basics of pump operation and venting and also knows terms such as *output variation, load variation*, and *torque* without having them defined. Passage B uses a minimum of technical language and explains the venting action by means of the simple analogy of the soda straw.

Guideline. Manuals for professionals may safely use more technical language and visuals, but must be as clear and logical as manuals for the general public.

General Public

Far more common are manuals for the general public (sometimes called consumer manuals). The term *general public* seems at first glance to be so broad that it defies definition. Begin by looking at the list below, which gives an idea of the categories of products usually intended for the general public user.

Typical General Public (Consumer) Products

1. Appliances
2. Automotive products
3. Biomedical devices (prostheses, blood pressure kits, contraceptives, heating pads, trusses, braces, contact lenses and glasses, orthopedic aids, hearing aids, dentures)
4. Construction equipment
5. Drugs and health products
6. Farm and industrial equipment

7. Firearms
8. Foods
9. Household products (soaps, polishes, cleaning agents, sprays, pesticides, stools, ladders)
10. Office equipment
11. Paints, general-purpose chemicals (fertilizers, solvents, removers)
12. Power and hand tools
13. Sporting goods (bicycles, minibikes, skis, swimming pools)
14. Toys

Despite the breadth of the term *general public*, the general public user may nevertheless be defined according to the following characteristics.

1. *Biological Characteristics.* Male or female, of any age

A general public user can be a male or female of any age, from child to elderly person. If you write instructions for skis designed for the resilient 16-year-old body, you have no assurance that a 40 year old won't try them. If you make a caustic toilet bowl cleaner for a housewife to use, you have no assurance that a preschooler (who can't read the label) won't think it's just another colored powder. Increasingly, many products formerly intended solely for one sex or the other are being used by both — such products as blenders, power saws, shotguns, and hair dryers.

2. *Literacy.* Illiterate to college educated

Illiteracy and declining literacy are modern realities. Instructions for use by the general public should not assume literacy. If the product is especially complex, unusual, or hazardous, reliance on visuals or pictures is essential, in the manual or in instructions affixed to the product. (The special problem of safety warnings is discussed in Chapter 6.)

3. *Technical Sophistication.* Sophisticated to naive

The general public user may be technically sophisticated or technically naive. If your product is intended for general use, you must aim the manual at the naive user. The technically sophisticated will be able to follow along anyway and will simply be able to use the manual more quickly. If the manual aims for the technically sophisticated user (and many manuals make this error), the naive or inexperienced user is left behind. The manual is then useless to that reader and goes unread.

Characteristics and Distinctions

Most manufacturers, by knowing the market for their products, can aim or target manuals, written instructions, and warnings by using the following list of user characteristics and distinctions. The list gives you more information for deciding if your users are professional or general public.

1. Personal characteristics of the user

   • Does he or she use this machine or product almost every day or only once in a while?
   • Is he or she likely to have or have used other products like it?
   • Does he or she do his or her own routine maintenance?
   • Does he or she do his or her own repairs? Should he or she?
   • Does he or she understand technical language?
   • Does he or she understand charts, circuit diagrams, mechanical drawings?

2. Conditions of manual use

   • Will he or she use the manual only to learn how to set up, use, or operate the product?
   • Will he or she refer to the manual or instructions often?
   • Will he or she use the manual or instructions only if something breaks, fails, or appears to be abnormal?
   • Will he or she read the entire manual or only a section here and there?
   • Will he or she be able to look at the machine or product when he or she is reading the manual?
   • Will the light be good?

3. Information wanted

   • Basic instructions for use, operation, and adjustment?
   • Routine maintenance procedures?
   • Sophisticated service procedures?
   • Specifications and parts lists?
   • Trouble-shooting procedures?
   • Explanations of new technologies or product features?

The same user may respond differently to different products. If as a writer you believe that you are writing too simply or too nontechnically, consider how the following situations demand simplicity and nontechnical clarity in written instructions:

- The housewife trying a new cooking oil? Repairing a faucet? Replacing spark plugs? Using a power saw?
- A 16-year-old boy repairing his bike? Making baby food?
- An 80-year-old man learning to use a computer? A power mower? A microwave oven?
- The 50-year-old independent auto mechanic servicing an older car? Servicing a late model car with parts formerly mechanically controlled, now computer controlled?
- A maintenance supervisor of heating and air-conditioning equipment fixing an air conditioner in a 30-year-old plant? Maintaining new equipment that uses new technology and corrosive chemicals?

None of these situations is impossible or even improbable. A professional in one situation can easily be a novice in another.

Guideline. General public manuals should be simple, clear, and nontechnical.

Guideline. Whenever the spectrum of users could range from sophisticated to naive, make manuals as nontechnical as possible.

Questions as Organizers of the Manual

After you have established a clearer idea of who is supposed to use your manual, you are ready to determine what the manual should contain and how its major sections should be organized. Ask yourself, "What is the manual supposed to do?" Your user can be your guide here. Lay out the manual sections as if they were responses to user questions. To do this, you must step back from your product and try to see it as first-time buyers or users might. What questions will they ask?

- The computer keyboard looks a lot like my typewriter. What's different about it? How do I underline? Indent? Set the line spacing? What else can it do besides type?
- My boss tells me this new paper-making machine is one of a kind. It looks like our old one to me. What's new? How much retraining will my foremen need to keep it running right?
- Interesting-looking toy—how does it work?

- Sunglo — I heard about this on TV. Let's see what's in it to make the windows shine. Do I have to mix it with water?
- What are the installation space dimensions for this? Must it be fireproofed?
- Crazy-looking copy machine. Where do you put the paper? How do you stop it?
- I could get this done a lot faster if I used my dad's power saw. How do you feed the board into it?
- Is it dangerous to use my electric razor while I'm sitting in the bathtub?
- If we buy this pump, what kind of retrofit is needed on our old equipment?

Imagine the buyers of your product talking to your industry vendors or shopping in a store or a dealer showroom. Buyers look at the product, talk to salespeople and vendors, or read the sales literature and instruction manuals. They do this because they have questions. Answers to those questions can form the major sections of your manual.

Look at this chart to see how a user question can form a section of your manual. The column on the left gives you a typical user question and its corresponding manual section. The columns to the right contain typical answers for two products, an exterior paint and a heavy duty wrecker. These answers could be used to make up the major sections of the manual.

Manual Section Derived from User Question	Answer to User Question about the Product	
	Exterior Paint	*Heavy Duty Wrecker*
Scope (*What is the main function of this product?*)	One of 25 products for protective exterior coating	Mounted on truck chassis; used for towing, lifting heavy vehicles
Description of Product (*Is this what I'm looking for? Introduce me to it.*)	Oil base, high quality, for residential or commercial buildings; 15 colors available	Boom and 2 winches; 2 telescoping outriggers on upright mast

Manual Section Derived from User Question	Answer to User Question about the Product	
Theory of Operation or Intended Use (How does it work? What is it for?)	Protective coating for wood, asbestos, brick stucco	Recovery operations using winches Choice of pulling or towing Towing: By front wheels By rear wheels
Special Feature or Design Details (What is special about it?)	Durability, nonfade color, controlled replacement color formula	Boom ratings: Extended: 12 tons Retracted: 35 tons Winch ratings: Safe load: 17½ tons
Limits of Operation or Use (What are its limitations?)	Not for metal, glass, plastic Brush application only—not spray Mix only with organic solvents; no water	Ratings apply only if: Truck chassis is adequate Both winches are attached to load Boom is at 15° from horizontal Load is lifted vertically
Setting up/ Turning on (How can I assemble it? Turn it on?)	Brush application Temperatures above 50°C	Wrecker installation on truck chassis requires special training
Normal Operation or Use (What is normal use and life of product?)	Dries in 24 hr 7-year life Can be washed	Heavily dependent on good maintenance and variations in weather conditions
Turning off/ Disposal (How do I turn it off? Dispose of it?)	Excess and accompanying solvents flammable Precautions for handling	Turn-off controls

Manual Section Derived from User Question	Answer to User Question about the Product	
Abnormal Operation (*What tells me something has malfunctioned?*)	Causes of cracking, peeling Not for internal use	Signs of malfunction Safety features Damage to cable, boom, or to load being towed or lifted
Preventive Maintenance (*How do I take care of it?*)	Proper surface preparation Proper application Close lid tightly	Complex machine Separate maintenance and repair manual
Storage (*How do I store it?*)	Store upside-down Shelf life	
Safety (*How do I use it safely?*)	Safety information will be found throughout the manual and on the product (See Chapter 6, Safety)	

Users as Feedback Sources

As we have shown, the manual user is important as the audience and organizer of your manual. The manual user can also be helpful in providing feedback, especially at the revision and follow-up stages of manual production. Writers who rely on user feedback tell us that this feedback step is invaluable for debugging a manual before the final copy is printed and for assessing manual use and effectiveness after the product is sold.

Simulating the User: the Person on the Street

How do you do that simulation? We have found that company efforts to collect user feedback range from informal and occasional to formal and systematic. Here are some of the techniques companies use:

1. Informal and in-house

   Some companies invite employees from other divisions, secretaries, friends, or family members — anyone who is a true stranger to the product — to "walk through" the manual, following its instructions and descriptions. Writers stand by and listen and watch, but they don't provide verbal backup to the manual unless the user asks for help. Wherever writers have to break in, explain more fully, or provide more information, the manual probably needs revision or clarification.

2. Beta-site testing

   Some companies, including those doing military contract work, have selected sites (beta sites) used for testing products and debugging manuals. At beta sites, users are asked to perform the operations described in the manual. For instance, an airplane mechanic may be asked to follow the manual for installation of a new landing gear. The mechanic is selected and identified as a typical user, a serviceman trained to work on military aircraft. All difficulties and snags experienced by the user in following instructions are monitored and recorded. The manual is corrected, revised, and retested to assure that instructions are clear. In the private sector, potential or long-time customers are asked to be beta sites for product and manual testing. Writers travel to the beta sites to observe the product and the prototype manuals being used by beta site employees. Companies who agree to be beta sites are usually compensated, by reduced price or some form of compensatory service. However, for many military contracts, beta-site testing is mandatory.

3. Protocol analysis

   As people perform tasks using a manual, they are often doing several things at once — reading, using their hands and/or feet, and talking, either to themselves or someone else. Protocol analysis watches what people do as they read and work, but it also listens. The manual users are asked to talk out loud, describing what they are doing or thinking. This verbalization of tasks is a rich source of information for manual producers. A user may stop, look puzzled, and say, "What screw? Where is the slot for it?" or mutter, "Abort? Error? Retry? I did what this thing told me to. How do I get out of this? It keeps saying Error, Error. Help!"

4. Formal systematic analysis

Some companies dedicate considerable money and time to formal user analysis. The setups can be as simple as the use of a hidden camera or videotape, plus an audiotape, used to record the manual user at work. More elaborate user feedback setups place users at a desk equipped with a microphone, the manual, and videotape monitors. Manual pages are presented and as users work their way through tasks, their activities and voices are recorded. Design engineers, safety engineers, and technical publications staff monitor the users at work from a studio equipped with one-way glass. The videotape record of the user at work is digitally synchronized with a videotape of the manual pages. The combination of voice plus video of both user and manual page allows observers to know exactly what the users saw on the page, what task they were trying to perform, and what they said and did about it.

User Interviews and Surveys

Many companies now do follow-up surveys and field checks on their manuals by asking users for feedback. Some companies simply include a postage-paid card in their manuals asking users to assess manual effectiveness. (Response in this way is likely to be low unless the manual is very bad.)

Some companies now use manual hotlines, listing a toll-free phone number in the manual and inviting people to call if they have trouble using the manual. Other companies make up a list of preferred, repeat customers and actively seek their feedback on manuals by making personal phone calls to customer-users or visits to sites where new products have been installed. These "personal touch" interviews, whether by phone or direct visit, are carefully planned surveys where specific questions are prepared ahead of time and feedback can then be used to improve manuals.

Manipulating Manual Elements to Match User Needs

The information you gather about your users should be your guide as you make choices in writing the manual. User analysis affects:

- Language level and reading level
- Choice and execution of visual and graphics
- Proportion of visual to verbal text

- Arrangement of segments of the manual
- Safety warnings
- Revision and updates

These elements may be manipulated in a number of ways to match user needs and questions. (See subsequent chapters for fuller treatment of visuals, graphics, and safety warnings.) Suppose, for example, that you are considering how to manipulate language levels and reading levels to accommodate users you have identified as either general public or professional. Example 2.2 shows how manual writers have controlled reading and language level.

Example 2.2. Two Passages Showing How User Analysis Affects Language and Reading Level. (Reprinted from *Dodge Dart, Coronet and Charger Service Manual 1967* (Detroit, MI: Dodge Division, Chrysler Motors Corporation, 1967), pp. 5–6. With permission; Muir, J. and Gregg, T. *How to Keep Your Volkswagen Alive: A Manual of Step by Step Procedures for the Compleat Idiot* (Santa Fe, NM: John Muir Publications, 1974), p. 5–10. With permission.)

Below are two different descriptions of the same procedure (removing front brake shoes from standard drum-type automotive brakes). The first is from a Dodge service manual; the second is from *How to Keep Your Volkswagen Alive: A Manual of Step by Step Procedures for the Compleat Idiot.*

From the Dodge Manual

1. Using Tool C-3785 remove secondary return spring then remove adjusting cable eye from anchor. (Note how secondary spring overlaps primary spring.)
2. Remove primary return spring.
3. Remove brake shoe retainer springs by inserting a small punch into center of spring and, while holding backing plate retainer clip, press in, and disconnect spring. Unhook cable from lever. Remove cable and cable guide.
4. Disconnect lever spring from lever and disengage from shoe web. Remove spring and lever.
5. Remove primary and secondary brake shoe assemblies and adjusting star wheel from support. Install wheel cylinder clamps (Tool C416) to hold pistons in cylinders.[2]

Continued

Example 2.2. Continued

From the Muir Book

    Look at the brake assembly; you'll find there are return springs holding both ends of the shoes toward each other. Look at how they're fastened — see how they're anchored. You'll have to replace them the same way. Clamp the vice [sic] grips on the closest spring to you then pry it out of its hole with the screwdriver. Remove the other springs the same. Take out the two round springs with caps over them in the center of the brake shoe webs, holding the shoes to the brake plate. They'll come off with your fingers, so hold the pin in the back of one with your forefinger and push and twist on the little cap with the thumb and forefinger of the other hand. When the little cap is 90° around on the pin, it will come off and the whole thing will come apart. Take a close look, you have to put them back on . . . Remove the other spring-cap-pin assembly from the other side. Now you can work them out of their slots. Snap a rubber band tight around the wheel cylinder slots so the wheel cylinders won't come apart.[3]

Comment. The procedures described in these two passages are identical, but the writers had two very different kinds of users in mind. Strengths and weaknesses of the two approaches include the following:

Dodge Manual

- Written for a professional mechanic
- Assumes familiarity with tools, parts, and mechanisms
- Numbered list format makes it easy to follow

Muir Book

- Written for novice mechanic, a do-it-yourselfer
- Anticipates trouble spots (especially those requiring unusual coordination of hand or hand and eye)
- Warns of disassembly problems
- Tries to use common terms for tools and parts
- Harder to follow than Dodge manual because of paragraph format instead of list
- Chattiness makes it wordy, but still more appropriate for novice

Feedback from users can also help you to make adjustments in manuals by revising and updating or by altering manuals as successive models of a product come out. Example 2.3 shows how user feedback from a survey can sharpen insights on actual use of the manual.

Example 2.3. Feedback from Manual Users: Surveys.

Talking with manual users, either informally or through structured surveys, can be informative and can give you insights on the strengths and weaknesses of your manuals.

To see what kind of feedback might be provided by manual users, we conducted a series of informal interviews with service people, dealers, and buyers of agricultural equipment. To encourage honest responses, we tried to ask questions that would encourage people to talk about areas of the manuals that most concerned them. We asked, "Do you use the manuals for your equipment? If so, how? If not, why not?"

We found that dealers and service people answer the questions somewhat differently than buyers (farmers). Here are some of the responses to our questions. Notice, as you read, that dealers and service people stressed the importance of the manual as a teaching tool and a legal document and that they were keenly aware of different levels of technical sophistication among buyers. Farmers had occasional positive things to say, but had many complaints about manual effectiveness. At the end of the user responses, you will find our comment on what we learned from this feedback exercise.

Informal Survey on Manuals for Agricultural Equipment Dealers and Service People

1. *Levels of Sophistication*

   - "There's a big difference in some of these modern-day farmers. There'll always be some who think the manuals are too complex — they want to be told to tighten down a nut, not what its dimensions are. On the other hand, if they're used to checking on a corn planter just by eyeballing it and then they buy one with electronic equipment, they'll sit down and read every word."
   - "No use lecturing a farmer about stuff he's used all his life. He tends to think all engines are pretty much the same. He knows how to drive a tractor; he's been doing it all his life. He'll look

at the manual if there's a new wrinkle, but he has to have it pointed out to him."

- "I guess there's a lot of the 'good old American know-how' in most of my buyers. I get my calls when they've taken something apart or tried to fix it and can't. If it looks like they've botched a job and the labor bill is going to get bigger if they go on with it, they'll call us to come out to bring it in."

- "Some of my buyers are buying for as many as 32 farms. They've got full-service departments to take care of their equipment—some of their mechanics are better than ours. They're very well educated, and they don't want to be talked down to."

- "The ones that need the most elementary help are these 90-day-wonder suburbanite farmers with 15 acres of land. They're probably buying a small tractor for the first time—everything's new to them."

- "You could mount the safety instructions in neon lights on the power takeoff, and 50 percent of the old-hand farmers will remove whatever gets in their way."

2. *The Importance of the Manual*

- "It used to be that the operator manual got filed away or tossed somewhere. Now they're getting more and more use. The equipment is more complicated, more things can go wrong, more malfunctions are adjustment problems."

- "Our service and delivery people use the manuals for teaching. In fact, we set up a school for all our buyers of corn planters, bailers, and combines. The service people run those schools, and they show the buyers what kinds of adjustments and troubleshooting are likely to present problems. They use the manuals to show a farmer what he can do for himself before he calls for help."

- "We think the manual is so important that we require dealers to register the serial number of the manual as well as the equipment.

1) We want to make sure he's got it and have proof of it.
2) We want to emphasize its importance to our customer.
3) Liability settlements are getting bigger all the time—there are more and more ways to get hurt."

- "One of the biggest complaints is that the manuals don't go deep enough. The pictures are fuzzy, or they don't know what kind of tool they need."

Farmers

3. *Mixed Responses*

- "The pictures are terrible. Half of them are too small and full of those damn little numbers. You can't see what the arrows are pointing at, and you have to keep flipping back and forth. Most farmers I know have eyesight that's none too good — all that crap flying around in the air, you know."
- "You ought to talk to about 500 farmers. I'll bet they'd all tell you those things are put together by a bunch of engineers sitting in offices with fluorescent lights. Most farmers aren't engineers. They're working in mud or in a dark tool barn."
- "I weld a lot. I know there's some stuff I shouldn't 'cause when I try to turn on something, I find out what I welded doesn't work so good anymore — wasn't made to be welded. But some of my stuff is pretty old, and a trip to the dealer is 35 miles and a five-hour ordeal, and then I might not even get the part. Who's got that kind of time?"
- "I wish they'd always give you the manual that goes with your equipment. Some of them don't really match the machine you've got. I can't tell you how many times I've stared at those pictures and wondered if that funny little thing sticking out of the picture is the same as the funny little thing on my machine."

Comment. Given a chance to talk about how they used manuals, these farmers, dealers, and service people had good and bad things to say about them. The most common responses tended to fall into the categories you see, although questions and conversations themselves were informal.

From these responses, we learned that

- Users are general public (range from technically sophisticated to technically naive).
- Manuals for products with new design features or repeated service problems get more use.
- Confusing visuals are a problem for many users.
- Users sometimes are given the wrong manual or a generic manual intended to serve for several models of a product.

- Dealers and service people rely heavily on manuals and sometimes use them for teaching.
- Conditions for manual use are sometimes bad (field, mud, barn).
- Farmers tend to use manuals for doing breakdown, maintenance, repair, and setup or for understanding new design features.
- Farmers work under pressure and do much of their own service and repair.
- Farmers probably seldom read the whole manual for a product they are familiar with.

Checklist: User Characteristics and Distinctions

Answer these user questions about the manuals you and your company produce. After you have answered the questions, consider whether you would want to change anything about the way your manual is put together.

1. Personal characteristics of the user

   ☐ Does he or she use this machine or product almost every day or only once in a while?
   ☐ Is he or she likely to have or have used other products like it?
   ☐ Does he or she do his or her own routine maintenance?
   ☐ Does he or she do his or her own repairs? Should he or she?
   ☐ Does he or she understand technical language?
   ☐ Does he or she understand charts, circuit diagrams, mechanical drawings?

2. Conditions of manual use

   ☐ Will he or she use the manual only to learn how to set up, use, or operate the product?
   ☐ Will he or she refer to the manual or instructions often?
   ☐ Will he or she use the manual or instructions only if something breaks, fails, or appears to be abnormal?
   ☐ Will he or she read the entire manual or only a section here and there?
   ☐ Will he or she be able to look at the machine or product when he or she is reading the manual?
   ☐ Will the light be good?

3. Information wanted

□ Basic instructions for use, operation, and adjustment?
□ Routine maintenance procedures?
□ Sophisticated service procedures?
□ Specifications and parts lists?
□ Troubleshooting procedures?
□ Explanations of new technologies or product features?

Checklist: User Questions as Organizers of the Manual

This checklist can be used either to plan and outline a new manual or to evaluate a manual you have already written.[4]

	User Questions	Segment of Your Manual That Answers Questions
1. Scope	1.	1.
2. Description	2.	2.
3. Theory of operation	3.	3.
4. Design detail	4.	4.
5. Limits of operation	5.	5.
6. Setting up and turning on	6.	6.
7. Normal operation	7.	7.
8. Turning off	8.	8.
9. Abnormal operation	9.	9.
10. Preventive maintenance	10.	10.
11. Storage	11.	11.
12. Safety	12.	12.

Adapted from Ranous, C.A., University of Wisconsin, Madison. With permission.

Summary

Be aware of the manual user as you plan your manual. The following steps will sharpen your awareness of the manual user and help to organize the manual.

- Think systematically about your manual user.
- Anticipate user questions.
- Employ answers to user questions to structure the manual.
- Employ user feedback to revise, refine, and update manuals.

References

1. Operators Manual, Model 8000, CooperVision Systems, an operating unit of CooperVision Surgical, a division of CooperVision, Inc., Cavitron Surgical Systems, Irvine, CA.
2. *Dodge Dart, Coronet and Charger Service Manual 1967.*(Detroit, MI: Dodge Division, Chrysler Motors Corporation, 1967), pp. 5–6.
3. Muir, J. and Gregg, T. *How to Keep Your Volkswagen Alive: A Manual of Step by Step Procedures for the Compleat Idiot* (Santa Fe, NM: John Muir Publications, 1974), p. 157.
4. Ranous, C. A. Checklist developed at the University of Wisconsin, Madison.

3

Organization and Writing Strategies

Overview

People tend to acquaint themselves with an unfamiliar product in much the same manner as they find their way around a new city. They look at maps and search for landmarks and street signs. As you develop a manual, you can simulate the function of street signs and landmarks by careful organization and by writing strategies that help readers find their way around the manual and feel comfortable with the product.

This chapter shows how to organize and write manual sections, which may be several pages, a cluster of several paragraphs, a single paragraph, or a single sentence in length. The following are the most commonly used organizing strategies.

1. General headings, followed by one or a number of pages that fall under that heading
2. Paragraph clusters, usually appearing as subheads under a general heading
3. Single paragraphs organized by the presence of a heading, a core sentence, and/or supporting sentences
4. Single sentences organized by various strategies, such as lists, parallels, series, comparisons, cause-and-effect statements

Organization of the manual also requires that you pay attention to the sequences of segments. We discuss approaches to sequencing: general-to-specific chronology and spatial logic. The last part of the chapter shows strategies to reduce the length of verbal text. Throughout the chapter, you will see examples of manual pages and, at the end of the chapter, some sample pages and questions that can be used for review.

General Headings: Chapters and/or Major Sections

We use the term *general headings* to denote those headings used for identification of manual chapters and/or major sections. General headings provide landmarks for user understanding of major systems of a machine, major uses of product, and/or major steps in a process. As suggested in Chapter 2, you can employ user questions in determining the major sections of a manual. We stated that those major sections would include:

Scope
Description
Operation and Intended Use
Special Features
Limits of Operation
Setting Up and Turning On
Normal Operation
Turning Off and Disposal
Abnormal Operation
Preventive Maintenance
Storage
Safety (throughout the manual)

Most manuals will have some, if not all, of these sections, and you may choose to vary the number, order, and length of major sections to match the requirements of your product. The following are some examples of the general headings used to designate chapters or major sections of manuals for a variety of products.

Examples of General Headings

Surgical Aspirator
Scope and Purpose of Manual
Introduction to Model 923
Description and Function of
  Handpieces
Description and Function of
  Console
Unipaks and Setup

Fabric Dye (Home Use)
Fabrics Suitable for Dye
Preparation of Fabric
Mixing and Preparing Dye
Applying the Dye
Fabric Care and Washing

Herbicide
Precautions
Uses of Product
Mixing and Spraying
Fluid Test Compatibility
Application Information
Cultivation Information
Weed Control

Tractor
Safety
Controls and Instruments
Pre-start Checks
Operating the Tractor
Drawbar and PTO
Ballast
Transporting
Wheels, Tires, Treads
Lights and Signals
Fuels and Lubricants
Lubrication and Maintance
Service
Storage
Trouble-shooting
Lubrication Chart

Table Saw
Safety
Unpacking and Cleaning
Connecting Saw to Power Source
Controls and Adjustments
Operation
Maintenance
Parts, Service Warranty

Camping Tent
Capacity of Tent and Users
Parts of Tent
Assembly
Disassembly
Care of Tent
Storage

The examples show both chronology and spatial logic and, of course, reflect the complexity of the product. For instance, the herbicide manual is a small, pocket-size booklet of 45 pages, the tractor manual is 102 pages, and the tent manual is 8 pages. As the length of the manual increases, the general headings become more important as road signs for the user, and for you, the writer, the general headings are invaluable as you develop the outline for your manual.

Headings as Hierarchical Cues

Headings not only tell the reader the main idea of the section to follow, but also provide cues on the relative importance of various kinds of information. These cues are largely picked up by responses to print size, type face, uppercase or lowercase letters, and position of the heading on the page. Individuals may disagree here and there on the meanings of these cues, but, in general, these responses hold true:

1. Largest level of organization or most important information

   • Larger type
   • Bolder print
   • All uppercase letters
   • Centered on the page

2. Smaller level of organization

   • Smaller type
   • Finer print
   • Mix of uppercase and lowercase letters, or all lowercase
   • Flush left or indented

3. Special emphasis

   • Italics or underline
   • Boxes
   • Color

For example, a section of a manual laid out in the following way will be perceived by the reader to be arranged in a descending hierarchy of ideas:

Controls

On-Off
Pre-set
Throttle
Turning

   • Right
   • Left
   • Circle

Choosing the Cues

If you are having your manual produced by professional printers or by in-house typesetting, you have many levels of hierarchy available to choose from (typefaces; print size; uppercase and lowercase). However, even if you are producing your manuals simply by typing them in the office, you still have the choice of all uppercase, all lowercase, or a mixture of upper and lower, plus underlining. If you set up all your headings in one mode only (such as uppercase), you are wasting valuable design options that are available merely by varying the choice of modes.

Desktop publishing seems, at first glance, to be the ideal intermediate solution. It saves on outside contract work and it places in the writers' hands a tempting array of choices, many more than those available on a conventional typewriter. Keep in mind, however, that page design is an art. The professional printer or typesetter brings to the task the visual expertise of the professional. With desktop publishing, you will have to learn this art yourself. The increase in flexibility and choice provided by desktop publishing is both bane and blessing. You have many more ways to "play" with page design, but doing so effectively takes time for development of new skills. Consider the tradeoffs before you decide to do everything yourself. (Chapter 4 discusses computer-assisted publishing.)

Look at Figure 3.1, which shows the power of print size, uppercase and lowercase letters, boxes, and color. When manual users see a page like Figure 3.1, they perceive the following hierarchies:

1. General Heading: STORAGE

   - The fifth major section of the manual.
   - Separated from the bottom text by full-page underscore line.
   - "Storage" is repeated at the bottom right of the page in italics on all subsequent pages that belong to this major section.
   - Very large "outline" type indicates further that "Storage" is a major section.

2. Subheads: STORING THE TRACTOR, STORING BATTERIES, REMOVAL FROM STORAGE

   - Subhead type is sized somewhat smaller than general heading, but larger than the type size for subsystems that fall under each subhead.
   - All uppercase letters reinforce hierarchical importance.

3. Subsystems: "Preparation," "Storing"

   - Combination of uppercase and lowercase letters plus still smaller print size indicates lower hierarchical level.

4. Boldface: Use for Subheads and Subsystems

   - Reinforces distinction between headings and the text proper.

Figure 3.1. Effective Use of General Headings. This figure shows the power of print size, uppercase and lowercase letters, boxes, and color. (Reprinted from *Service Manual, Series 2 Four-Wheel Drive Tractors,* Applicability: 1977 Production (Canada, Winnipeg, Manitoba, Versatile Manufacturing Ltd., 1977), p. 55. With permission.)

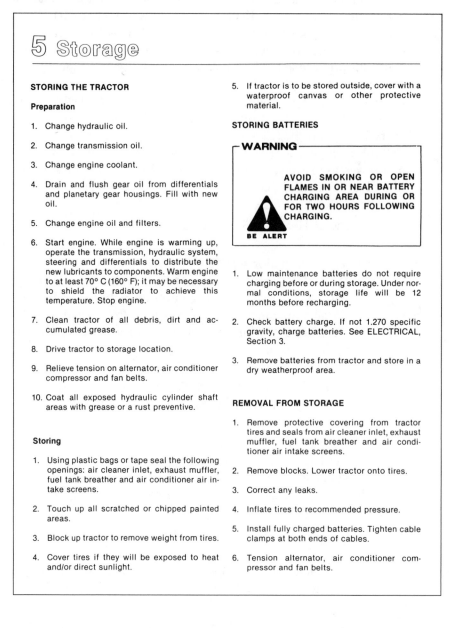

5 Storage

STORING THE TRACTOR

Preparation

1. Change hydraulic oil.

2. Change transmission oil.

3. Change engine coolant.

4. Drain and flush gear oil from differentials and planetary gear housings. Fill with new oil.

5. Change engine oil and filters.

6. Start engine. While engine is warming up, operate the transmission, hydraulic system, steering and differentials to distribute the new lubricants to components. Warm engine to at least 70° C (160° F); it may be necessary to shield the radiator to achieve this temperature. Stop engine.

7. Clean tractor of all debris, dirt and accumulated grease.

8. Drive tractor to storage location.

9. Relieve tension on alternator, air conditioner compressor and fan belts.

10. Coat all exposed hydraulic cylinder shaft areas with grease or a rust preventive.

Storing

1. Using plastic bags or tape seal the following openings: air cleaner inlet, exhaust muffler, fuel tank breather and air conditioner air intake screens.

2. Touch up all scratched or chipped painted areas.

3. Block up tractor to remove weight from tires.

4. Cover tires if they will be exposed to heat and/or direct sunlight.

5. If tractor is to be stored outside, cover with a waterproof canvas or other protective material.

STORING BATTERIES

WARNING

AVOID SMOKING OR OPEN FLAMES IN OR NEAR BATTERY CHARGING AREA DURING OR FOR TWO HOURS FOLLOWING CHARGING.

BE ALERT

1. Low maintenance batteries do not require charging before or during storage. Under normal conditions, storage life will be 12 months before recharging.

2. Check battery charge. If not 1.270 specific gravity, charge batteries. See ELECTRICAL, Section 3.

3. Remove batteries from tractor and store in a dry weatherproof area.

REMOVAL FROM STORAGE

1. Remove protective covering from tractor tires and seals from air cleaner inlet, exhaust muffler, fuel tank breather and air conditioner air intake screens.

2. Remove blocks. Lower tractor onto tires.

3. Correct any leaks.

4. Inflate tires to recommended pressure.

5. Install fully charged batteries. Tighten cable clamps at both ends of cables.

6. Tension alternator, air conditioner compressor and fan belts.

5. Boxes: Used for Safety Warnings

   • Boxes plus safety alert symbol in color call special attention to fire hazard.

Paragraph Clusters, Subheads, Core Sentences

Hierarchy of ideas can also be suggested by the way you handle paragraph clusters. For example, a mechanism or process may have a number of subsystems or subroutines that you want to treat as a unit in a paragraph cluster. Subheads and core sentences are useful devices to keep paragraph clusters focused and to organize and highlight key ideas of the paragraph(s). Core sentences serve as forecasting devices for what follows in the rest of the paragraph or section. Look at Example 3.1.

Example 3.1 Paragraph Cluster with Subheads and Core Sentences.

OPERATION OF THE SPECTROPHOTOMETER
   To operate the Ace Model X Spectrophotometer, three procedures must be followed. These include preliminary steps, placement of samples in the cell compartment, and measurement of samples.

Preliminary Steps

   Before making any measurements, turn the sensitivity switch to standby, the shutter switch to SHTR, and the power and lamp switches to the ON position. Then allow the machine to warm up for 15 minutes before attempting any measurements . . .

Placement of Samples in Cell Compartment

   Each machine has two rectangular-shaped test tubes called cells. One of these cells contains a reference sample. This sample . . .

Measurement of the Samples

   Slide the cell holder into position so that the reference cell is in the light path of the compartment. Select the desired wavelength . . .

Comment. Example 3.1 shows how the initial (core) sentences set up expectations of subsequent development of the paragraph cluster. The word *three* in sentence 1 and the list of three steps in sentence 2 are then picked up and repeated, in the same order, in the subheads that follow. Common sense principles of organization like these are simple to apply, but easy to forget in the rush of putting together a manual. For the reader, they are the road signs that help the reader feel, "Okay, I'm still with you. I'm not getting lost."

Syntax Strategies

Elements in single sentences or groups of sentences offer additional opportunities for writers to provide road signs for the reader. Among the most useful for manual writers are lists, parallels, series, and comparison-contrast.

List Strategy

When steps or sequences are being described, the list strategy is one of your most valuable tools. Many manuals rely too heavily on the linear mode, stringing out a series of instructions horizontally in long, complicated sentences and paragraphs. The horizontal arrangement makes instructions more difficult to follow. For example, if a novice photographer is developing film for the first time in a home darkroom, he or she will be following a series of steps very closely. Seeing those steps numbered and arranged vertically helps the user to keep his or her place. The user tracks the process by thinking, "Step 1 finished. Now for step 2."

The list strategy makes use of a powerful communication tool, the vertical. To understand how it works, add the following set: 456 + 1678 + 45 + 789 + 9. Keep track of how long it takes you to do this. Now add:

$$
\begin{array}{r}
357 \\
4789 \\
23 \\
540 \\
8 \\
\hline
\end{array}
$$

The number of units, tens, hundreds, and thousands is the same in both sets, but most people are far quicker with the vertical. The list is the verbal equivalent of the numerical set. In Example 3.2 the words *before* and *when* serve as predictable organizers, like units and tens, and the reader's eye picks up only the key words as it sweeps down the passage. Note also that the white space around each element makes the list even clearer.

Example 3.2. Linear Mode vs. List Strategy.

The following paragraph is written in *linear* mode:

The system must be vented under the following circumstances:

Before starting an engine that has not been operated for an extended period of time. When the fuel filters have been replaced. When an engine, in operation, runs out of fuel. When any connections between the injection pump and fuel tank have been loosened or broken for any reason.

Rewritten, its focus is sharpened by the *list*:

The system must be vented under the following circumstances:

- Before starting an engine that has not been operated for an extended period of time
- When fuel filters have been replaced
- When an engine, in operation, runs out of fuel
- When any connections between injection pump and fuel tank have been loosened or broken

Parallels

When procedures or mechanisms are closely related, parallel sentence strategies also serve to sharpen focus. Notice the difference in clarity in the following examples:

Nonparallel

A. Install front bolts with the threads down. On the rear bolts, make sure the threads face up.

Parallel

B. Install front bolts with threads down.
   Install rear bolts with threads up.

Readers can understand instruction **A**, but they will grasp **B** more quickly.

Series

If you are listing a series of parts or steps, keep the series grammatically consistent:

Inconsistent verb forms

A. Always wait for the tractor to come to a complete stop. After lowering the equipment to the ground, make sure the transmission is shifted to the N position; the park brake should be set to prevent the tractor from rolling. Then remove the key.

Consistent verb forms

B. Always wait for the tractor to come to a complete stop, then lower equipment to the ground, shift the transmission to N position, set the park brake so the tractor will not roll, and remove key.

Readers will understand **A**, but they will have to slog through shifts from active to passive voice and unnecessary glitches in verb tense sequence ("wait," "after lowering," "is shifted," "should be set," "remove"). In **B**, each element of the series begins with an imperative verb ("wait," "lower," "shift," "set," "remove"). The series of commands is predictable, comfortable—in short, a clear map of what to do.

Comparison-Contrast

If you are trying to make comparisons or contrasts between steps or characteristics of a process or product, use parallel structure to heighten the comparison:

Nonparallel comparison

A. More modern mechanisms have electronic controls, whereas they were formerly operated manually.

Parallel comparison

B. Modern mechanisms are electronically controlled, whereas formerly mechanisms were manually controlled.

The first example is comprehensible, but the comparison is not sharp.

Sequencing

General to Specific

We have said that users respond to unfamiliar products and their manuals by searching for landmarks. Another way to think about this is to consider how you respond to meeting a stranger. You begin with a general impression of a person strange to you and only later begin to notice and understand details of dress, manner, or speech. The principle is a simple one—it is psychologically more natural to move from general to specific. Knowing this, you should try to introduce your user to the product by arranging the material of the manual from general to particular.

One way to assure this progression from general to specific is to think of your manual in terms of *overview* first, then *details*. Suppose, for example, that you want to provide instructions for a stereo system. The first section of the manual should contain an overview (a labeled, full-shot photo or drawing, a verbal description, or both) of the entire system: tuner, amplifier, tape deck, turntable, and speakers. This overview gives the reader a sense of how the parts relate to each other. Then by means of details (subsystems), you may begin to provide more detailed information for each of the parts.

This same general-to-specific strategy also holds true for descriptions of processes or procedures. For example, before you begin to describe the individual chemicals and processes in a home permanent wave, tell the reader, by means of an overview, that the process will take well over an hour and include several subset processes: testing the hair for reaction to chemicals; preparing the hair (trimming, washing, etc.); applying curling lotion and curlers; rinsing and neutralizing; shampooing; setting the hair; caring for the hair after the permanent. The overview can be brief— a list will do—but it sets up a forecast or expectation of the subroutines to follow.

Use the following guide as a way of ensuring a general-to-specific arrangement:

Overview — Introductory paragraphs or sections, with accompanying overview photo or drawing explaining the system, machine, process, or product

Details — Components of system, mechanism, or procedure

Spatial

For very large products, such as industrial machines, trucks, tractors, cranes, and motorcycles, users say they approach the product with a preconceived spatial logic. For example, they may think of the product from front to back, from top to bottom, or from left to right when facing the product. Of course, variations of this basic perception will be required in specialized manuals or service instructions. The logical view for service manuals on exhaust systems would be to show and describe the system as seen from below by the mechanic as he or she works with the system overhead on a hoist.

Whatever the product or process, it helps to ask your person-on-the-street user, "How do you think of this? Do you stand in front or at the side? Do you think of this from front to back? Top to bottom?" Then arrange the manual to match the user's spatial perception. Do not, for example, begin with the rear axle assembly on a truck or tell the user of a sewing machine how to embroider or buttonhole before you explain how to thread the machine or sew a straight seam.

Chronology

Most processes and procedures have an inherent chronology, i.e., the steps for doing something grow naturally out of the way the product or process works. For example, the user will usually want to know about setup or assembly before operating procedures, maintenance, or storage.

However, at the level of paragraph or subsection, exceptions to strict chronology are quite common. Suppose that you are describing the operation and use of a home whirlpool. The dangers posed by high water temperatures to the elderly and to people with heart conditions or high blood pressure need to be mentioned, both in the manual and on the product, *before* the owner uses the whirlpool. Likewise, you would not instruct a user how to remove the cover of a pressure cooker without first showing how to make sure that all the steam has escaped from the cooker. In short, if any step or procedure can, in its execution, cause damage to the product or injury to the operator, be sure to explain this before the step is listed.

Anticipate trouble spots in procedures, even if chronology is interrupted. Do not warn of troubles or dangers when it is too late.

Example 3.3. Maintaining Proper Chronology.

Finishing Your Home-Built Furniture

Advice too late

1. Prepare wood surface by using fine sandpaper or steel wool.
2. Mix oil-based stain of your choice.
3. Apply stain thinly with brush or cloth.
4. Allow to stand 5 minutes.
5. Remove stain by wiping.
6. Allow stained surface to dry for 24 hours.
7. Apply final finish of urethane or varnish.

Note: Some soft woods absorb stain quickly. If you are unsure of your wood, apply stain to test area first.

Advice on time

Many oil-based stains work best on hardwoods like cherry, walnut, or oak. If your furniture is made of soft wood, such as pine, test a sample area first to check absorbency of stain. If stain soaks or smears, prime surface with thinned shellac before finishing. (Thinned shellac: 2 parts solvent/1 part shellac)

1. Prepare wood surface . . .
2. Test sample area for stain absorption.
3. Apply thinned shellac primer if necessary and allow to dry thoroughly.
4. Apply stain thinly . . .

Patterns for Describing Mechanisms and Processes

Some writers find it useful to work from a rough organizational outline to ensure a general-to-specific arrangement of manual segments. Below are two outlines that you may use as guidelines. Be flexible in using them to make sure they match the special features of your product.

Pattern for a Description of a Mechanism

1. Introduction

   a. Definition
   b. Purpose
   c. General description
   d. Division of a device into its components

2. Principle or theory of operation

   a. Divisions—what the part is, its purpose, and its appearance
   b. Divisions into subparts

      (i)   Purpose
      (ii) Appearance—often through visuals
      (iii) Details—shape, size, relationship to other parts, connections, material

3. The operation of the system

   a. Ways in which each division achieves its purpose
   b. Causes and effects of the device in operation

Pattern for a Description of a Process

1. Introduction

   a. General information as to why, where, when, by whom, and in what way the process is performed or occurs
   b. List of the main steps
   c. List of the components involved

2. Description of the steps or analysis of the action

   a. First main step (or sequence of events)

      (i) Definition
      (ii) Special materials

   b. Division in substeps

3. Conclusion (summary statement about the purpose, operation, and evaluation of the whole process)

Combining Strategies

We have described a number of writing and organizational strategies:

- General headings (sections or chapters)
- Paragraph clusters (with subheads and core sentences)
- Syntax strategies (lists, series, parallels, comparison contrast)
- Sequencing strategies (general to specific, spatial, chronological)
- Patterns of describing mechanisms and processes

These strategies seldom occur in isolation; rather, writers tend to combine the techniques to ensure smooth flow and internally consistent logic. Look at Figures 3.2, 3.3, 3.4, 3.5, and 3.6 to see how writers have combined a number of strategies.

Figure 3.2. Combined Writing Strategies: General to Particular, Paragraph Clusters, Chronology. The sequence of pages follows the natural order of unpacking, cleaning and assembly. Paragraph clusters are placed close to relevant visuals. Instructions are present tense, active verbs. (Reprinted from *10" Motorized Table Saw Instructor Manual, Part No. 340213* (Pittsburgh, PA: Delta International Machinery Corporation, 1986), pp. 4−7. With permission.)

UNPACKING AND CLEANING

Carefully unpack the saw, stand and all loose items from the carton. Remove the protective coating from the saw table surface. This coating may be removed with a soft cloth moistened with kerosene (do not use acetone, gasoline or lacquer thinner for this purpose). After cleaning, cover the table surface with a good quality paste wax.

ASSEMBLY INSTRUCTIONS

ASSEMBLING STAND

Assemble the lower shelf (A) and the two top tie bars (B) to the two side panels (C), as shown in Fig. 2, using the sixteen 3/8″ x 5/8″ long carriage bolts (eight of which are shown at (D) Fig. 2), sixteen 3/8″ flat washers and sixteen 3/8″ hex nuts.

Fig. 2

Figure 3.2. (continued)

ASSEMBLING RUBBER FEET TO STAND

Turn the stand upside down and assemble the four rubber feet (A) Fig. 3, to the bottom of the two side panels using the four 3/4" long hex head screws (B), 1/4" flat washers (C) and 1/4" hex nuts (D). After the rubber feet are assembled, return the stand to the upright position.

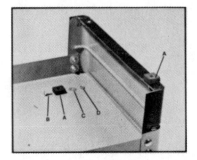

Fig. 3

ASSEMBLING BLADE GUARD SUPPORT ROD

1.   Make certain the saw is disconnected from the power source.

2.   Turn the saw upside down and place it on a non-scratch surface.

3.   Position threaded end of blade guard support rod (A) Fig. 4, through hole (B) in back of saw cabinet.

4.   Insert opposite end of blade guard support rod (A) Fig. 4 (the end with a flat on it) into hole in trunnion (C), and tighten screw (D).

Fig. 4

ASSEMBLING SAW TO STAND

Position the saw (A) on the stand (B), as shown in Fig. 5. Line up the four holes on the bottom of the saw cabinet with the four holes on the top of the stand and fasten the saw to the stand using the four 5/8" long hex head screws, 3/8" flat washers and 3/8" hex nuts.

Fig. 5

Figure 3.2. (continued)

ASSEMBLING BLADE RAISING AND TILTING HANDWHEELS

1. Assemble the blade raising handwheel (A) Fig. 6, to the blade raising screw (B) making sure the slots (C) in the hub of the handwheel are engaged with the roll pins (D) on the raising screw shaft.

Fig. 6

2. Screw lock knob (E) Fig. 7, on end of raising screw shaft.

3. Assemble tilting screw handwheel (F) and lock knob (G) Fig. 7, to the blade tilting screw shaft in the same manner, as shown in Fig. 7.

Fig. 7

ASSEMBLING EXTENSION WINGS

1. Assemble extension wing (A) Fig. 8, to the saw table using the three screws and washers (B). With a straight edge (C) Fig. 9, make sure the extension wing is level with the saw table before tightening the three screws (B) Fig. 8.

2. Assemble the other extension wing to the opposite end of the table in the same manner.

Fig. 8

Fig. 9

Figure 3.2. (continued)

ASSEMBLING SAW BLADE

1. Make certain the saw is disconnected from the power source.

2. Remove the table insert (A) Fig. 10.

3. Raise the saw blade arbor (B) Fig. 10, to its maximum height by turning the blade raising handwheel counter-clockwise and remove the arbor nut (E) and flange (D) from the saw arbor.

4. Assemble the saw blade (C) to the saw arbor making sure the teeth of the blade point down at the front of the table, as shown in Fig. 10, and assemble the flange (D) and arbor nut (E) to the saw arbor and tighten arbor nut (E) as far as possible by hand, being sure that the saw blade is against the inner blade flange.

Fig. 10

5. Using the open end wrench (F) Fig. 10 and Fig. 11 supplied, place the wrench (F) on the flats on the saw arbor to keep the arbor from turning and tighten arbor nut (E) using the remaining wrench (G) Fig. 11, by turning the nut counterclockwise.

6. Replace table insert (A) Fig. 11, making certain that it is flush with table surface.

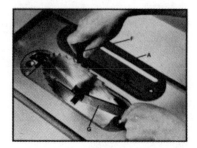

Fig. 11

ASSEMBLING GUIDE RAILS

1. The guide rail (A) Fig. 12, with graduations is to be assembled to the front of the saw table with the gradua-tions up.

2. Insert the two special screws (B) Fig. 12, through the two holes (C) in the guide rail and place spacers (D) between the guide rail (A) and saw table. Thread the two special screws (B) into the tapped holes in the saw table. Do not completely tighten the two screws (B) at this time.

3. Insert special screw (E) Fig. 12, through hole (F) in guide rail and place spacer (G) between guide rail and extension wing. Fasten with washer and nut (H). Tighten three screws (B) and (E) to fasten guide rail to table and extension wing.

4. Assemble the remaining guide rail to the rear of the table in the same manner.

Fig. 12

Figure 3.3. Combined Strategies: Clusters, Bold Face, Glossary-List. This "older" version of the manual uses a vertical format. Reader will look at diagram, then at callout, and then at explanation. Segmenting of long lists into groups of five makes them less forbidding to read and keep track of. (Reprinted from *Pierce Chassis Operators Manual* (Appleton, WI: Pierce Manufacturing, Inc.), pp. 9–10. With permission.)

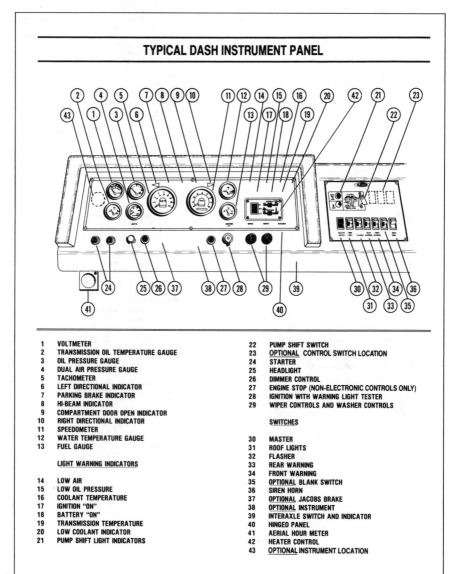

TYPICAL DASH INSTRUMENT PANEL

1 VOLTMETER	22 PUMP SHIFT SWITCH
2 TRANSMISSION OIL TEMPERATURE GAUGE	23 OPTIONAL CONTROL SWITCH LOCATION
3 OIL PRESSURE GAUGE	24 STARTER
4 DUAL AIR PRESSURE GAUGE	25 HEADLIGHT
5 TACHOMETER	26 DIMMER CONTROL
6 LEFT DIRECTIONAL INDICATOR	27 ENGINE STOP (NON-ELECTRONIC CONTROLS ONLY)
7 PARKING BRAKE INDICATOR	28 IGNITION WITH WARNING LIGHT TESTER
8 HI-BEAM INDICATOR	29 WIPER CONTROLS AND WASHER CONTROLS
9 COMPARTMENT DOOR OPEN INDICATOR	
10 RIGHT DIRECTIONAL INDICATOR	**SWITCHES**
11 SPEEDOMETER	
12 WATER TEMPERATURE GAUGE	30 MASTER
13 FUEL GAUGE	31 ROOF LIGHTS
	32 FLASHER
LIGHT WARNING INDICATORS	33 REAR WARNING
	34 FRONT WARNING
14 LOW AIR	35 OPTIONAL BLANK SWITCH
15 LOW OIL PRESSURE	36 SIREN HORN
16 COOLANT TEMPERATURE	37 OPTIONAL JACOBS BRAKE
17 IGNITION "ON"	38 OPTIONAL INSTRUMENT
18 BATTERY "ON"	39 INTERAXLE SWITCH AND INDICATOR
19 TRANSMISSION TEMPERATURE	40 HINGED PANEL
20 LOW COOLANT INDICATOR	41 AERIAL HOUR METER
21 PUMP SHIFT LIGHT INDICATORS	42 HEATER CONTROL
	43 OPTIONAL INSTRUMENT LOCATION

Figure 3.3. (continued)

TYPICAL DASH INSTRUMENTS AND CONTROLS

NO.	IDENTIFICATION	NORMAL USE OR READING
1	Voltmeter	Indicates battery condition and rate of charge or discharge.
2	Main Transmission Oil Temperature Gauge	Registers main transmission oil temperature. Normal operating range is 160°-220°F (71°-104°C). Maximum is 250°F (121°C).
3	Oil Pressure Gauge	Indicates engine oil pressure in PSI. Stop engine immediately if low or no pressure is indicated. (5-70 PSI)
4	Dual Air Pressure Gauge	Air pressure should be from 90-120 PSI while operating.
5	Tachometer	Indicates engine speed (RPM).
6	Left Turn Signal Indicator	Flashes green when left turn signal is ON.
7	Parking Brake Indicator	Illuminates red when parking brake is set.
8	Hi-Beam Indicator Light	Illuminates blue when headlights are on high beam.
9	Compartment Door Open Indicator	Illuminates when door(s) is(are) open.
10	Right Turn Signal Indicator	Flashes green when right turn signal is ON.
11	Speedometer/Odometer	Indicates vehicle speed and records total accumulated mileage.
12	Water Temperature Gauge	Indicates cooling system temperature. (170°-195°F, 205°F; 77°-88°C, 96°C).
13	Fuel Gauge	Indicates level of fuel in tank. Fill fuel tank at the end of each day's operation to prevent condensation.
14	Low Air Indicator	Illuminates when air pressure is low.
15	Low Oil Pressure Indicator	Illuminates when oil pressure is low.
16	Coolant Temperature Indicator	Illuminates when cooling temperature is high.
17	Ignition Switch Indicator	Illuminates green when ignition switch is ON.
18	Battery ON Indicator	Illuminates green when battery switch is ON.
19	Transmission Temperature Indicator	Illuminates when oil temperature is high.
20	Low Coolant Indicator	Indicates loss of engine coolant.
21	Pump Shift Indicator Lights	Green: OK to pump. Red-flashing: Pump not in gear.
22	Pump Shift Switch	To shift pump in or out of gear.
23	Optional Switch Location	May be used for optional equipment.
24	Start Buttons	Push black buttons to start engine.
25	Headlight Switch	Pull ON push OFF for headlight control.
26	Dimmer Control	Turn to vary instrument light intensity.
27	Engine Stop (Non-Electronic Controls Only)	Push red button to stop engine (Non-electronic controls only).
28	Ignition Switch/Warning Light Tester	ON/OFF switch for engine electrical power. Turn switch completely clockwise.
29	Wiper Controls	Wiper—turn knob clockwise for ON and push to wash.
30	Master Switch	ON/OFF rocker type for electrical power.
31	Roof Light Switch	ON/OFF rocker type—push top for ON.
32	Flasher Switch	ON/OFF rocker type—push top for ON.
33	Rear Warning Switch	ON/OFF rocker type—push top for ON.
34	Front Warning Switch	ON/OFF rocker type—push top for ON.
35	Optional Blank Switch	ON/OFF rocker type—push top for ON.
36	Siren Horn	ON/ON rocker type—push top for siren; push bottom for horn.
37	Optional Jacobs Brake	Control location.
38	Optional Instrument	May be used for optional instrument.
39	Interaxle Switch and Indicator	Control location.
40	Hinged Gauge Panel	To access instrument panel for service.
41	Hour Meter (Aerial)	Records aerial hours of operation.
42	Heater Controls	Heater/Defroster temperature and fan control.
43	Optional Instrument Location	May be used for optional instrument

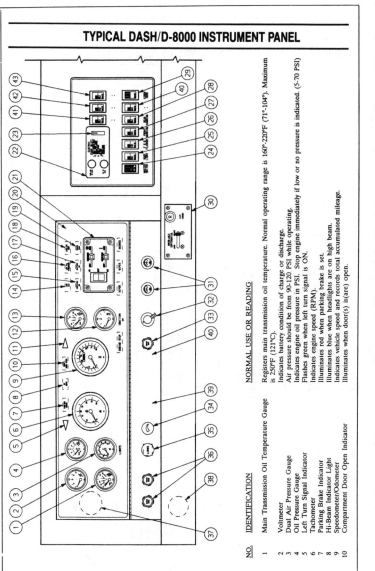

TYPICAL DASH/D-8000 INSTRUMENT PANEL

NO.	IDENTIFICATION	NORMAL USE OR READING
1	Main Transmission Oil Temperature Gauge	Registers main transmission oil temperature. Normal operating range is 160°-220°F (71°-104°). Maximum is 250°F (121°C).
2	Voltmeter	Indicates battery condition of charge or discharge.
3	Dual Air Pressure Gauge	Air pressure should be from 90-120 PSI while operating.
4	Oil Pressure Gauge	Indicates engine oil pressure in PSI. Stop engine immediately if low or no pressure is indicated. (5-70 PSI)
5	Left Turn Signal Indicator	Flashes green when left turn signal is ON.
6	Tachometer	Indicates engine speed (RPM).
7	Parking Brake Indicator	Illuminates red when parking brake is set.
8	Hi-Beam Indicator Light	Illuminates blue when headlights are on high beam.
9	Speedometer/Odometer	Indicates vehicle speed and records total accumulated mileage.
10	Compartment Door Open Indicator	Illuminates when door(s) is(are) open.

Figure 3.4. Revised Version of Pages Shown in Figure 3.3. Uses clusters, glossary-list, and horizontal format. In revising, the writer reduced the cross-reference job for the reader. Reader will now look at diagram and callout—two steps, rather than three. Horizontal format improves sequencing and alignment of numbers. (Reprinted from *Pierce Chassis Operators Manual* (Appleton, WI: Pierce Manufacturing, Inc., 1990), pp. 9–10. With permission.)

TYPICAL DASH/D-8000 INSTRUMENTS AND CONTROLS

#		
11	Right Turn Signal Indicator	Flashes green when right turn signal is ON.
12	Fuel Gauge	Indicates level of fuel in tank. Fill fuel tank at the end of each day's operation to prevent condensation.
13	Water Temperature Gauge	Indicates cooling system temperature. (170°-195°F, 205°F; 77°-88°C, 96°C).
14	Low Air Indicator	Illuminates when air pressure is low.
15	Ignition Switch Indicator	Illuminates green when ignition switch is ON.
16	Battery ON Indicator	Illuminates green when battery switch is ON.
17	Low Oil Pressure Indicator	Illuminates when oil pressure is low.
18	Coolant Temperature Indicator	Illuminates when cooling temperature is high.
19	Transmission Temperature Indicator	Illuminates when oil temperature is high.
20	Low Coolant Indicator	Indicates loss of engine coolant.
21	Heater Controls	Heater/Defroster temperature and fan control.
22	Pump Shift Indicator Lights	Green: OK to pump. Red-flashing: Pump not in gear.
23	Pump Shift Switch	To shift pump in or out of gear.
24	Master Switch	ON/OFF rocker type for electrical power.
25	Roof Light Switch	ON/OFF rocker type-push top for ON.
26	Flasher Switch	ON/OFF rocker type-push top for ON.
27	Front Warning Switch	ON/OFF rocker type-push top for ON.
28	Rear Warning Switch	ON/OFF rocker type-push top for ON.
29	Siren Horn	ON/ON rocker type-push top for siren; push top for siren; push for horn.
30	Interaxle Switch and Indicator	Control location.
31	Wiper Controls	Wiper-turn knob clockwise for ON and push to wash.
32	Ignition Switch/Warning Light Tester	ON/OFF switch for engine electrical power. Turn switch completely clockwise.
33	Engine Stop (Non-Electronic Controls Only)	Push red button to stop engine (Non-electronic controls only).
34	Dimmer Control	Turn to vary instrument light intensity.
35	Headlight Switch	Pull ON push OFF for headlight control.
36	Start Buttons	Push black buttons to start engine.
37	Optional Instrument	
38	Optional Instrument	
39	Optional Jacobs Brake	
40	Optional Switch Location	Control location.
41	Optional Switch Location	
42	Optional Switch Location	
43	Optional Switch Location	

NOTE: Gauge Panel is hinged to access instrument panel for service.

Figure 3.4. (continued)

Figure 3.5. Combined Strategies: General to Particular, Bold Face List, Glossary.
Visuals and component lists on first page provide overview. Next page provides
details for one component, the control box. (Reprinted from *Service Manual for
Modular Nugget Ice Maker Model MH 750* (Vernon Hills, IL: Scotsman Ice
Systems), pp. 10–11. With permission.)

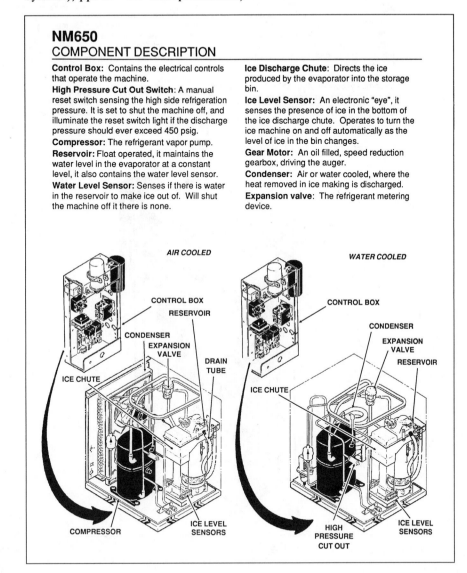

NM650
COMPONENT DESCRIPTION

Control Box: Contains the electrical controls that operate the machine.

High Pressure Cut Out Switch: A manual reset switch sensing the high side refrigeration pressure. It is set to shut the machine off, and illuminate the reset switch light if the discharge pressure should ever exceed 450 psig.

Compressor: The refrigerant vapor pump.

Reservoir: Float operated, it maintains the water level in the evaporator at a constant level, it also contains the water level sensor.

Water Level Sensor: Senses if there is water in the reservoir to make ice out of. Will shut the machine off it there is none.

Ice Discharge Chute: Directs the ice produced by the evaporator into the storage bin.

Ice Level Sensor: An electronic "eye", it senses the presence of ice in the bottom of the ice discharge chute. Operates to turn the ice machine on and off automatically as the level of ice in the bin changes.

Gear Motor: An oil filled, speed reduction gearbox, driving the auger.

Condenser: Air or water cooled, where the heat removed in ice making is discharged.

Expansion valve: The refrigerant metering device.

Figure 3.5. (continued)

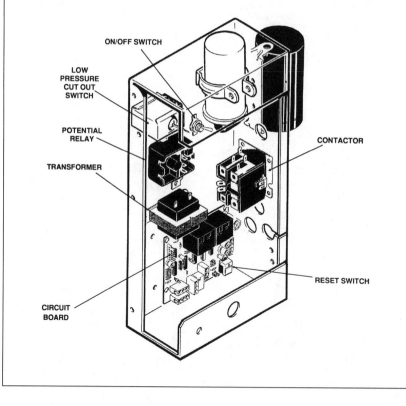

NM650

COMPONENT DESCRIPTION: Control Box

Contactor: A definite purpose contactor connecting the compressor and the remote condenser fan motor to the power supply.

Circuit Board: Controlling the ice machine through sensors and relays. The sensors are for ice level and water level. The relays are for the gear motor (with a built in time delay to clear the evaporator of ice when the unit turns off) and for the compressor. The reset switch is mounted on the circuit board.

Transformer: Supplies low voltage to the circuit board.

Low Pressure Cut Out Switch: A manual reset control that shuts off the ice machine when the low side pressure drops below a preset point, 0-4 psig.

Potential Relay: The compressor start relay.

On/Off Switch: Manual control for the machine.

Reset Switch: Part of Circuit Board, manual reset. Lights up when unit shuts off from: ice discharge chute being overfilled (opening the microswitch at the top of the chute); low or high pressure switches opening.

ON/OFF SWITCH

LOW PRESSURE CUT OUT SWITCH

POTENTIAL RELAY

TRANSFORMER

CONTACTOR

RESET SWITCH

CIRCUIT BOARD

Figure 3.6. Combined Strategies. Products with keyboards, control panels, and graphics use glossary format for parts identification. (Reprinted from *Model M1117A Multichannel Thermal Array Recorder* (Waltham, MA: Hewlett-Packard, 1989), p. 1–9. With permission.)

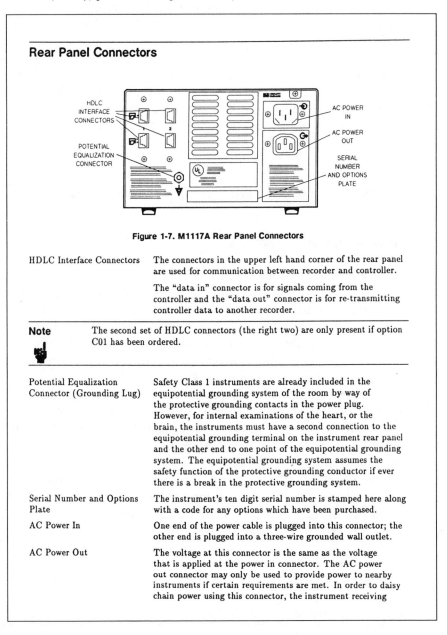

Rear Panel Connectors

Figure 1-7. M1117A Rear Panel Connectors

HDLC Interface Connectors	The connectors in the upper left hand corner of the rear panel are used for communication between recorder and controller.
	The "data in" connector is for signals coming from the controller and the "data out" connector is for re-transmitting controller data to another recorder.
Note	The second set of HDLC connectors (the right two) are only present if option C01 has been ordered.
Potential Equalization Connector (Grounding Lug)	Safety Class 1 instruments are already included in the equipotential grounding system of the room by way of the protective grounding contacts in the power plug. However, for internal examinations of the heart, or the brain, the instruments must have a second connection to the equipotential grounding terminal on the instrument rear panel and the other end to one point of the equipotential grounding system. The equipotential grounding system assumes the safety function of the protective grounding conductor if ever there is a break in the protective grounding system.
Serial Number and Options Plate	The instrument's ten digit serial number is stamped here along with a code for any options which have been purchased.
AC Power In	One end of the power cable is plugged into this connector; the other end is plugged into a three-wire grounded wall outlet.
AC Power Out	The voltage at this connector is the same as the voltage that is applied at the power in connector. The AC power out connector may only be used to provide power to nearby instruments if certain requirements are met. In order to daisy chain power using this connector, the instrument receiving

Strategies for Reducing the Length of Verbal Text

Manuals are seldom read under the relaxed conditions in which one might, for example, read a novel or a technical reference book. Because manual users are reading and dealing with the product at the same time (starting, adjusting, repairing), you should try to balance sections of solid prose, arranged in paragraphs, with other information-providing strategies. A number of writing strategies can be used to reduce the bulk of solid prose text. The most obvious of these are visuals (Chapter 5) and formatting (Chapter 4). Other techniques that serve to provide relief from too much print and to make manuals more quickly comprehensible are

- Illustrations and examples
- Omission of unnecessary theoretical/technical background

Illustrations and Examples

Illustrations and examples provide another way to reduce the length of written text. They can show alternative ways of carrying out a process or can distinguish between the right and wrong ways of doing things. If you look back at the strategies suggested in this chapter, you will see that they are supported by illustrations and examples. For instance, after you read a description of the list strategy, you were provided with an example showing conversion of a linear passage to a list.

Illustrations and examples may also be used in combination with written text as a shorthand to describe a procedure more briefly. In essence, what you are doing is showing users, rather than merely describing in words, how a product will look after a step or procedure has been completed. Illustrations and examples are especially useful for products whose operation involves many steps or procedures. Manuals for electronic devices, monitors, and computers used for graphics, numbers, or word processing all make use of examples and illustrations to clarify what the written text describes (see Figure 3.6).

Omission of Unnecessary Theoretical/Technical Background

Good organization of a manual means knowing not only what to include, but also what to leave out. Manual writers with technical backgrounds are often tempted to include too much theoretical material. The theory or experimental design that underlies a product will always be of interest to the manufacturer as well as to the technically sophisticated manual user. However, in manuals for the general public, you should be guided most of all by what the user needs to know.

Not only does excessive theoretical explanation take up page space and increase manual costs, it is also forbidding and distracting for naive manual users.

Some samples of excessive theoretical/technical material:

- A manual for a surface tension measurement device in which the first four pages describe the history of surface tension study, going all the way back to Archimedes
- A manual for an inexpensive stereo system that describes several decades of development work on printed circuits
- A manual for a sailboat that devotes over half its space to unresolved controversies on vector theory and wind motion

You must use your judgment to determine how much theory is appropriate for adequate understanding of your product, but remember that the average buyer of a microwave oven, for example, will probably not be interested in the esoteric physics of microwaves. Much of the material found in operator manuals for the general public would be more appropriate in the service manual (see Chapter 7).

Review of Effective Writing Strategies

Using Example 3.4, take a minute to review your understanding of the writing strategies covered in this chapter.

Comment. The two sets of instruction for the tape recorder-radio have the same information, but **B** is more effective than **A**. Passage **B** is easier to follow because:

- Sentences are shorter and begin with the active verbs.
- The variety of typefaces and use of boldface for names of controls helps the user locate controls more quickly on the tape recorder itself. The writer uses caps for all buttons (STOP, PLAY, EJECT, OFF) and boldface and initial caps for all parts of the player **(Cassette Door)**.
- Wording for controls exactly matches wording on the product.
- Numbered-list strategy is used by both writers, but list effectiveness in Passage **A** is blurred by muddy, strung-out sentence patterns.

Example 3.4. Review of Effective Writing Strategies: Comparison of Two Sets of Instructions.

Read these two sets of instructions for using a portable tape recorder-radio. How would you answer these questions?

1. Which set of instructions is easier to follow?
2. What writing strategies make that set easier?

Passage A

1. Before playing tapes, make sure the radio switch is turned off.
2. To open the cassette door, press the button that stops the player and ejects the tape.
3. Insert a cassette into position so that the tape side is facing upward and the tape itself is on the right. Then press the cassette into place securely.
4. Push the cassette door closed and press the play button.
5. If you want to adjust the sound, adjust the controls for volume level and tone.

Passage B

1. Set RADIO switch to OFF.
2. Press STOP/EJECT button to open the **Cassette Door**.
3. Insert cassette into position as shown — open tape side up and full reel toward the right. Press the back of the cassette all the way in.
4. Close the **Cassette Door** and press PLAY.
5. Adjust VOLUME and TONE for desired sound.

Summary

Intelligent use of organization and writing strategies

- Makes the structure of the manual "jump" off the page
- Provides important landmarks for the reader
- Reduces the bulk of the verbal text

Pay special attention to headings, core sentences, paragraph clusters, and syntax strategies and organize your material from general to specific.

Be careful in deciding how much material to include and minimize solid blocks of prose by the use of charts, tables, examples, and illustrations:

Tell users what they need to know.
Omit material that is merely nice to know.

4

Format and Mechanics

Overview

In the various chapters of this book, we discuss what kind of information and what kinds of supporting materials (photos, diagrams) to include in an instruction manual. In addition, we discuss how best to organize and write the manual. This chapter discusses how to arrange this material so that readers can easily find their way through it to the particular information they need. The best writing in the world is wasted if the pages are visually unappealing (users won't read them), the referencing is inadequate (they won't be able to find what they need), or the mechanical elements—paper, binder, etc.—are inadequate (the manual will be too hard to use). Skillful design choices in these matters, however, will maximize the manual's usefulness and help ensure that the user will read and refer to it often.

Since the first edition of this book, the computer has transformed the workplace for most manual writers. The ubiquitous computer monitor has replaced the typewriter (and often, the drafting table), and more and more production work is performed in-house. This change has made the technical writer's job much broader: "writers" do not just write anymore; they also do page design, type speccing, graphics, and paste-up. Learning about the elements that contribute to the visual impression of a manual is even more important as decisions of format and mechanics fall more and more to the writer. We begin this chapter with a quick look at computer-assisted publishing.

What is Computer-Assisted Publishing?

In reality, computers have been used in the publishing effort for some time. Professional typesetting has been computerized for years—the compositor in the green eyeshade choosing type and locking it into the composing stick is gone with the buggy whip. What has changed is that typesetting and page design are no longer exclusively done by professionals: anybody with the right equipment can produce material in different typefaces, in multiple-column formats, and so on. And the right equipment has become remarkably inexpensive.

What is the right equipment? That depends on your needs. At the low end, all you need is a personal computer, a laser printer, and a sophisticated word processing program, all of which can be obtained for under $6000. At the high end, you might have dedicated workstations linked in a network that includes computer–aided design (CAD) stations, using very complex publishing software and producing camera-ready pages with color separations—for several tens of thousands of dollars.

At either extreme, the writer is in charge of aspects of manual production that used to be exclusively dealt with by specialists. Twenty years ago, writers typically wrote text longhand. Typists turned the words into readable text, graphic artists supplied illustrations, photographers provided photos, and all these raw materials went to a production manager. This individual sent the text to be typeset, after choosing fonts and sizes, sent the photos to be screened (again after choosing sizes), and designed the page layout. When the typeset material and screened photos came back, a paste-up artist literally cut and pasted, producing camera-ready pages. Different people were responsible for different aspects of production, and each of these people was a specialist with skills and competence in a particular area.

With a computer-assisted publishing system, the writer can do it all. The writer can choose type fonts and sizes, create illustrations with "draw" or "paint" software, scan photos into computer memory format, and lay out pages using software designed for that purpose. The results can be close to professional publishing quality or just awful, depending on the degree to which the writer has developed the necessary graphic and design skills.

Computer-assisted publishing appears to be here to stay, and that means that writers must pay attention to all aspects of page design—not just writing. This book is not the place to provide a full course in the principles of page design or to go into detail about the choices available in publishing software and hardware. As you begin to make decisions about the look of your manuals, keep these two rules in mind:

1. *Remember the goal: Make it easy for the reader.*

   Design your pages so that they look accessible and inviting and so that the reader can easily find the information he or she needs.

2. *Keep it simple.*

   Computer-aided publishing offers you a dazzling array of choices of type size and style, borders, column layout, and so on. Simple is usually best. Remember, in manuals, you want the reader to focus on the information, not the appearance.

Format/Layout

Strictly speaking, *format* refers to the mechanical specifications of the page — page size, column width, type size, etc. — and *layout* refers to the placement of actual text (copy) and visuals (art) on the page. In this book, since we are giving general guidelines, we will use *format* to cover both areas — both have to do with the visual impression a page gives. This visual impression is remarkably important. All of us have picked up a manual or report, glanced through it, and thought, "This looks too hard to bother with" — before we had actually read any of it. On the other hand, we have also picked up articles or reports and thought, "This looks interesting. I think I'll read it" — again, before we had actually read any of it. The difference in the two responses is largely a result of the visual appeal of the page. In the first case, perhaps the type was too small to read easily or was set in one unbroken block, with no white space for eye relief. In the second case, perhaps eye-catching headings provided us with a sense of what was covered before we started reading — giving us a head start on understanding.

The writers of instruction manuals, more than many kinds of technical writers, must be conscious of this visual appeal. Managers will read technical reports because they must — they need the information. They may be frustrated by bad format, but they will still make an attempt to plow through the document. The user of a product, however, is all too likely to ignore the manual if it looks hard to use — ignore it, that is, until something breaks because he or she did not read the instructions and did not use the product properly.

There are four elements to work with in designing an effective format: the shape of the text, type size and style, headers and footers, text headings, and white space. We will look at each of these elements separately, although they must work in concert for the format to be effective.

Shape of the Text

Look at a page and draw an imaginary line around a block of text. This is what we mean by the shape of the text. It can be perfectly rectangular or irregular along the right-hand side; it can fill up the whole page or be one of several blocks on that page. In typesetting terms, the shape of the text is determined by column width, column depth, and the contour of the right-hand margin — i.e., whether the margins are *justified* (even, like a newspaper column) or *ragged right* (like the right-hand margin of a typed page). If you work for a large company, the decisions about column width and so forth may already be determined for you. Most companies require that all their manuals follow a standard format. If you work for a small company, or if your company is just getting into the business of operator manuals, you may be called upon to make these decisions yourself (see also Chapter 1, Planning). Here are some general guidelines to help you. (Even if you work in a company that requires a standard format, you may be able to improve on it.)

Column Width. Column width is a function of line length, which in turn is a function of the type you use. Readability studies have shown that the optimum line length is about an alphabet and a half to two alphabets or roughly 40 to 50 characters. With a line much shorter than that, the eye must keep jumping to the next line, and words and phrases are frequently broken, causing fatigue and decreased comprehension. If the line is much longer, the eye has too far to travel back to begin the next line and is likely to settle on the wrong line of type. (A corollary is that the longer the line, the more space is needed between lines.) Clearly, the smaller the type you use, the shorter will be a line composed of 40 characters. A normal typed page has a line length at least half again as long as the recommended length, but we are so used to it that readability seems not be affected. However, in operator manuals, where the reader is likely to be looking at the manual, then at the product (to locate a part or do a procedure), then back at the manual, a two-column format facilitates the reader's finding his or her place easily. Another common format that provides a good line length is the "2/5" format. In this format, a standard $8^{1}/_{2}$-in. × 11-in page is separated into two uneven columns: a left-hand column about 2 in. wide, used for headings and explanatory notes, and a right-hand column about 5 in. wide, used for text. If the type face is not too small, this format works well. Remember that with a two-column format you can still run a photo or a diagram the full width of the page if a one-column width would make it too small to read easily.

Column Length. Columns ordinarily run the full length of the page, although column length may be affected somewhat by the illustrations you use and by the way you use white space. (The use of white space is discussed in a subsequent section.)

Justified or Ragged Margins. In typeset or computer-published material it is possible to justify the right-hand margin, i.e., to make it even. This requires adjusting the spacing between words and letters to make all the lines come out the same length. It also ordinarily requires frequent hyphenation of words. There is no right or wrong decision about whether to justify the right-hand margin. Generally, justified copy is perceived as more formal and may be slightly harder to read. If you are using a desktop publishing system, check how well your system can adjust spacing for right-justified text. Some of the less sophisticated systems tend to adjust primarily word spacing—resulting in text that is very hard to read. In that situation, ragged right is definitely a better choice.

Type Size and Style

All the various type styles may be divided into two groups: *serif* type and *sans serif* type. Serifs are the little lines (sometimes only suggested) at the ends of each stroke in a letter. Serif type has these little lines, and sans serif does not (*sans* is French for "without"). Figure 4.1 shows examples of serif and sans serif type.

Figure 4.1. Samples of Serif (right) and Sans Serif (left) Type. Sans serif typefaces normally use strokes of single weight, or width. The Omega typeface shown here is unusual, in that the lines widen out toward the ends—almost suggesting the beginnings of serifs.

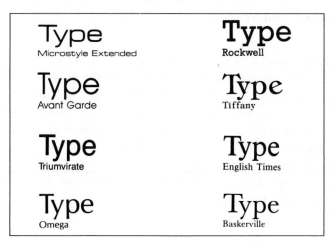

Sans serif type has a "modern" look to it, but readability studies indicate that most people perceive it as harder to read. No one is quite sure why this is true—it may be that the serifs tend to "seat" the letters on the line and pull the eye along to the next word. Sans serif type is fine for headings, but for the text proper, we recommend using a serif type.

Similarly, a mixture of uppercase and lowercase letters is easier to read than text written in all uppercase. The reason for this is clear: lowercase letters show much more variation in shape than uppercase letters, permitting easier recognition.

Type size is measured in points, with 72 points to an inch. To understand what is meant when we say that something is set in, say, 12-point type, imagine a standard typewriter key. A little metal block with a raised letter on it hits the ribbon to make a letter appear on the page. In traditional typesetting, a page would be set by lining up a series of individual blocks with raised letters on them and locking these into a wooden frame to hold them in position. The point size of the type measures the height of the *block*—not just the letter itself. Thus, the letters in two different styles of 12-point type may be slightly different heights, but the blocks would be the same. Even though most typesetting today is done by computer, the old terms are still used.

You can see that the tallest letters in 12-point type would be a bit less than one sixth of an inch high. Figure 4.2 shows examples of several type sizes. Most text is set in 8- to 12-point type. The smallest size that can be read without a magnifying glass is 6-point type. Because operator manuals are likely to be read under less than ideal conditions, we would recommend going no smaller than 8-point type for anything in the manual and would urge using 10-point or 12-point type for ease of reading.

Headers and Footers

Headers (also known as "running heads") are headings at the top of a page that tell the reader what part of the book he or she is in. For example, it is common to show the chapter title at the top of the left-hand page and a subhead from within the text on the right-hand page—as is done in this book. Less commonly, this information is given at the bottom of the page instead of the top, in which case the labels are called footers (not, to our knowledge, "running feet"!). More commonly, footers, if used at all, are used for such information as revision number, series number, and so on.

Figure 4.2. Samples of Different Type Sizes, Measured in Points. For ease of reading, type smaller than 10- or 12-point is not recommended.

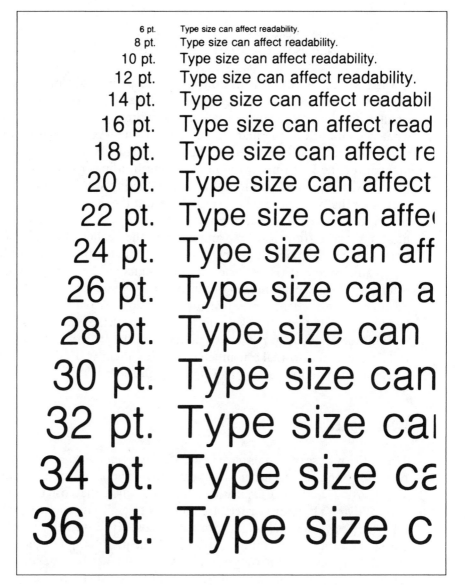

Headings

Headings are signposts that give the reader a sense of what is covered in a section of the manual (see also Chapter 3, Organizing and Writing Strategies). Most readers will skim over a section by reading just the headings before starting to read the details. For this reason you should make sure to:

- Use enough headings.
- Use headings that reveal the organization of the section.

Most writers use too few headings. Instead of just labeling the major sections of a chapter, consider adding subheadings to point out the smaller divisions. Remember that, as the writer, you are familiar with what is contained in the manual you are writing—you know where the parts are and what is covered in each. Your readers are looking at the manual for the first time; without sufficient headings to help them find their way, the manual will look like a bewildering sea of prose.

Be sure that your headings accurately reflect the organization of material in the manual. Use varying type sizes and boldface type to set the headings apart from the text and to give the reader clues about the structure. Thus, major divisions should be signalled by larger, boldface type, and minor divisions should be designated by correspondingly smaller type. One caution is in order: never use more than three levels of headings in text—it becomes just too confusing. Taken as a whole, the headings should form the skeleton, or outline, of the chapter. Not only will this technique help readers to understand the structure of the material, but it will also make it much easier for them to find a particular piece of information just by leafing through the book.

White Space

The judicious use of white space in an operator manual can improve both readability and comprehension and can provide your reader with another set of visual cues to organization. In times when the cost of paper is constantly rising, many writers are tempted to cram as much as possible onto each page in order to cut production cost. However, this practice is akin to cramming everything possible into one photograph or diagram—you end up with a confusing jumble that the reader won't even bother trying to decipher. We have all found ourselves put off by one page of text and attracted by another, largely as a result of an unconscious (or conscious!) assessment of how difficult it would be to read. An important factor in this judgment is white space.

Adequate white space can by itself improve readability simply because it makes it physically easier to read the page. Our eyes, like the rest of our bodies, need rest breaks. If these mini-rests are not provided to us in the form of white space, our eyes will take them anyway—and we will find ourselves having to reread a sentence because we missed a few words.

Also, most people are able to read, i.e., move their eyes along the line of text, faster than their minds can follow. White space gives the mind time to assimilate information before going on to the next piece.

We do not read word by word, piling individual words one on top of the other, but rather in chunks—adding phrases and sometimes whole sentences together to form the thought. For the mind to grasp a chunk of information, it must see it as a chunk, and white space can help with this. Even such a simple use as leaving a space between paragraphs makes a difference. It helps the mind to see the information in the paragraph as a unit.

White space can also give information about the structure of a piece of writing. In the same way that headings can show how the manual is organized, white space, if used carefully and consistently, can show how a section of text is put together. Thus, a blank line or two (such as between paragraphs) lets your reader know that you are moving from one unit to another of equal importance. Similarly, indenting a section (widening the white space around it) lets your reader know that you are moving to a smaller organizational division within a single unit. Surrounding an item with white space will also call attention to it, such as setting off a warning from ordinary text for emphasis.

Using white space to signal the organization of a chapter will work only if you use it consistently: e.g., if you always use the same number of blank lines between major divisions and always use the same (but smaller) number between minor divisions. Again, some of these decisions may already be made for you by company policy, but you may wish to go further than the minimum specifications of a standard format. You will find it easiest to be consistent if you make these decisions at the outline stage and make a table to help you remember what you decided. Otherwise, you will be having to page or scroll back through completed material to find whether to leave two lines or three between sections.

Careful use of white space may add a tiny bit to the cost of a manual because not every page is crammed corner to corner, but it will help ensure that the manual is used. If the manual sits untouched on the shelf, the whole cost of producing it was wasted.

Referencing

Because people rarely read operator manuals cover to cover like a novel, referencing is extremely important to help the reader find the section he or she needs. There are three types of referencing available for use in writing an operator manual: the table of contents, the index, and cross-referencing. Let's look at these in turn.

Table of Contents

The table of contents is essentially a map of the book. It appears in the front of the manual and outlines what the manual covers, giving page numbers of the beginning of each section. The reader will probably look there first for a specific item of information. Any manual more than a few pages long should include a table of contents.

To be useful, the table of contents must be arranged so that it clearly shows the organization of the manual and is easy to use. The names used to designate sections in the table of contents should match the headings used in the text. You may not want to include every tiny subhead in the table of contents, but those you do include should have the same wording in both places. You should also show organizational levels in the table of contents, perhaps by type size or by indention, to reflect the arrangement of information in the text. Be careful to make it readable. If the page numbers on the right are too far away from the words on the left, the reader may find it difficult to know what goes with what. You may wish to run dot leaders across the page or leave spaces between small groups of headings and page numbers. You must strike a balance between making the table of contents too skimpy and making it so complete that it nearly reproduces all the information in the manual itself.

In addition to (or instead of) the traditional table of contents, other means may be used to help a reader find his or her place. Some manuals use a tab system, whereby the sections of the manual are divided by tab pages to make it easy to flip to the section you need. Some use colored pages to distinguish one section from another. Some manuals use a two-level table of contents: a general one at the beginning and a specific one at the head of each section. Whatever method or combination of methods you choose, the same basic rules apply: make it logical, reflective of the organization of the manual itself, and easy to use.

Index

The index is an alphabetical listing of subjects and the numbers of the pages on which they appear. It is usually placed at the back of the manual and is more comprehensive and detailed than the table of contents. It does not, of course, show the structure of the manual, but it is useful for locating a specific item of information quickly.

Preparation of an index used to be a tedious process of putting each entry on a separate index card and going through the text writing down page numbers as each entry appeared. The use of computers has made it somewhat easier, since the computer can readily search for and locate words. However, developing a good index is never a mechanical process, for one needs to index concepts, not just words. The computer can help make sure that you do not miss any occurrences of a particular word, but a human still needs to make the decisions about what words to include. Keep in mind that the purpose of an index is to help readers find information: include words and concepts that you think someone might want to look up.

Cross-Referencing

Often a reader using an operator manual needs to be aware of information in a section other than the one he or she is using at the moment. For example, in a lawn mower manual, the section on winter storage of the mower might say to drain the oil and replace it before the next use. The location of the drain plug and the proper weight oil to use are listed in the section on maintenance. Rather than let the reader page through the manual at random looking for this information, point it out: "See 'Maintenance,' page 4." Put yourself once again in the reader's shoes and remember that he or she does not know the contents of the manual backward and forward as you do. Whenever you think it would be helpful for the reader to be referred to another section of the manual, do so. Of course, the most obvious place for cross-referencing is in the trouble-shooting section. This is usually set up as a table with headings like "Problem," "Probable Cause," and "Remedy." Too often, the remedies suggest something like "adjust spark plug gap" without telling the reader where in the manual the proper gap is given. To save your readers a lot of frustration, always include the full information needed—title of section *and* page number. If they are using the trouble-shooting section, they are frustrated enough already.

Of course, all of these suggestions will mean more changes when the manual is revised. For this reason, some companies, whose products undergo frequent model changes, instruct their writers to write "generically" and not include specific references. As with many aspects of manual design, what makes the manual easier for the reader makes it harder or more expensive for the company. Each company has to decide where the appropriate tradeoff lies.

In the matter of cross-references, computer technology can help. Some companies are working toward putting manual information into a database, with information cross-linked, so that page changes—and even specification changes—can be made automatically. Such a system requires a large investment of time up front, but it can streamline production in the long run.

Mechanical Elements

The mechanical elements of a manual include the paper it is printed on, the cover, the binding, and the size. You must choose these as carefully as you choose the photos to be included or the manual's organizational structure. These seemingly superficial elements may make or break a manual because they affect ease of use and durability. No one set of rules will cover all applications, but in this section we will present some factors to consider as you make your choices.

Paper

Two major questions arise about the paper used in manuals:

1. *How durable is it*? Is the paper easily torn? Is it flimsy and likely to shred with hard use? Clearly, if your manual will be referred to again and again, you must use durable paper. In addition, thin, flimsy paper may allow bleed-through, i.e., allow the print on one side of the page to be seen from the other side, which makes for harder reading.
2. *How porous is it*? Porosity will affect how the paper accepts ink and thus how sharply photos will reproduce. Porous paper will allow the ink to bleed, thus obscuring fine detail. Porous paper will also accept other substances, like oil and dirt. If your manual is likely to be used under dirty conditions, you should choose a harder-surfaced paper, possibly even a coated stock (although this is quite expensive).

Cover

The cover is both a mechanical element in the manual and a public relations device. Most companies have a standard cover format for their manuals, including information about the model and often a picture. This dual purpose dictates the questions you must ask yourself when you design a cover and choose the cover material.

1. *Is it attractive*? Will the user want to read it, and does it speak well for the company?
2. *Is it durable*? The manual cover protects the inside pages and must thus be made from heavier stock. If the manual is likely to be used in a harsh environment (rain, oil, etc.), it should be made of coated stock, possibly even of vinyl.

Binding

The purpose of the binding is to hold the pages together so that they can be easily read. Here are some factors to consider:

1. *Will the pages lie flat?* Nothing is more irritating than trying to do a procedure requiring both hands and frequent glances at the instructions, only to find that the manual flops shut each time you let go.
2. *Will the pages begin to fall out after hard use?* This is a common problem with "perfect" bindings, though rarely with stapled, stitched, or spiral bindings.
3. *Will frequent additions or corrections be sent to owners?* If the manual is likely to be updated often, it might be a good idea to put it in a ring binder, so that outdated pages can easily be replaced with new ones. If the binding requires that holes be punched in pages, make sure that margins are wide enough to allow this to happen without losing part of the text.

Size

The size of the manual is related to how it will be used. An $8^1/_2$-in. × 11-in. three-ring binder would be an awkward size for a manual for a 35 mm camera — you want something small to fit in a camera bag. On the other hand, $8^1/_2$-in. × 11-in. is a good size to use on a workbench in a garage. Think about how your customer will want to use the manual. Will he or she want to tuck it in a pocket or have it handy on a shelf? In general, small, oddly shaped manuals are easier to lose than more standard-size ones, but there are no hard-and-fast rules. As always, a clear understanding of your audience and of the manual's purpose will guide you to the right decision.

Summary

In this chapter we have discussed several aspects of manual construction that contribute to making the manual easy to use. Careful choice of format, including type size and style, headings, and the use of white space, can make manual pages attractive and easy to read. Providing adequate referencing in the form of tables of contents, indexing, and cross-referencing (especially in trouble-shooting sections) can make it easy for readers to find the information they need. Good design choices for the mechanics of the manual (page and cover stock, binding, and size), based on probable conditions of use, will help ensure that the manual you create will be read and referred to often. Care and attention given to these "surface" details will maximize the effect of the work you have put into writing, organizing, and creating good visuals.

5

Visuals

Overview

In all manuals, the visuals—photographs, drawings, charts, and tables—may be more important than the words. Clear, readable instructions and descriptions are necessary; clear visuals are vital. Most users, confronted with a manual for the first time, will leaf through the "pictures" long before they will read the words. The reader will see the pictures before the text and will usually expect them to be more accessible than the text. Most users will try to figure out a procedure from the visuals first and will read the text only as a last resort.

The visuals take on even more significance when the expected users do not read well or when the manual is written for an international market—both situations that increasingly confront the manual writer.

Regardless of language skills, all users will turn to visuals for two purposes:

- To identify parts
- To substitute for experience

Manual users will rely on the pictures to help make unfamiliar part names clear: when told to "tap leg closures firmly until well seated," the new barbecue grill owner will look at the drawing to find (with relief) that the leg closures are merely the little plastic caps that go over the ends of the grill's tubular metal legs. Parts identification is an important function of visuals that is at work throughout the manual, from the initial set-up chapter to the parts catalog at the end. Words absolutely cannot substitute for good visuals in helping the reader get to know the

product. Figure 5.1 shows a typic the pictu
identification.   easier to
                                    ıre than it
                                    , operator,
) demonstr

Figure 5.1. Typical Use of a Visual to Identifuals. Pictı
17291 Revision D, AMX-4 Operation (Milwate parts of t
Systems, 1989), p. 4. With permission.)   ough a proc
                                            ıt, or degre
                                            oridge only
                                            signed, poc
                                            image of
                                            few complai
                                            how to cho
                                            ımum effect

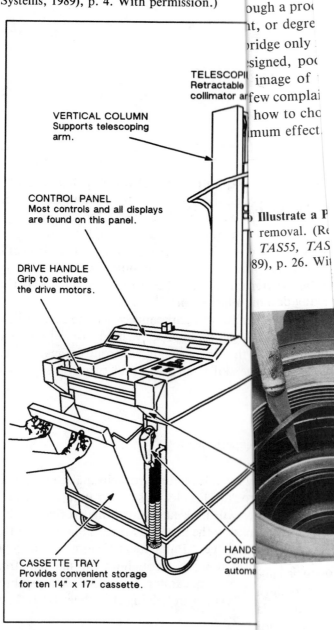

TELESCOPI
Retractable
collimator ar

VERTICAL COLUMN
Supports telescoping
arm.

CONTROL PANEL
Most controls and all displays
are found on this panel.

DRIVE HANDLE
Grip to activate
the drive motors.

CASSETTE TRAY
Provides convenient storage
for ten 14" x 17" cassette.

HANDS
Contro
automa

Illustrate a removal. (Re
, TAS55, TAS
89), p. 26. Wi

Figure 5
here he
Rotary
(Quincy
permiss

Compres

When to

First
uals u
the pr
engine
get th
your
about
the in:
mind.
the m
and
obscu

Photo

Photo
rangii
specia
paper
graph
the p
patter

will see shades of grey rather than individual dots. (The dot effect can easily be seen by looking at a newspaper photograph under a magnifying glass.) More recently, computer technology has enabled photographs to be digitized — or electronically broken into a dot pattern that can then be saved as a computer file.

Photos have a number of advantages over drawings (single-tone or line art) for manual writers. The most important of these is that they are easily understood by the most technically naive user. A photograph of a product or a part *looks like* that product or part. The user needs no special training to interpret the visual.

In addition, a photo generally shows a part in context, surrounded by other parts of the product. This can aid a new user in identifying unfamiliar parts. This realism carries with it two disadvantages.

First, photos may be cluttered and not isolate the important part. Second, they cannot be used for views of hidden areas — you cannot very easily have a cutaway photo! (In Chapter 7, Service Manuals, there is a cutaway view that looks like a photo. It is, however, an unusual example.) The problem of eliminating clutter and isolating the important part may be dealt with in several ways. Two of the most common ways are to use contrasting arrows or outlining to point out the important parts and to airbrush out background distractions (see Figures 5.4 and 5.5).

For other ideas, consult a commercial artist or photographer. The

Figure 5.4. Example of the Relevant Area of a Photo Outlined in a Contrasting Tone. This technique is useful for combining the realism of a photo with the ability of a drawing to focus the reader's attention. (Reprinted from *Operating Instructions*, IBM Correcting Selectric and IBM Selectric II, © 1973, International Business Machines Corporation, White Plains, NY, p. 6. With permission.)

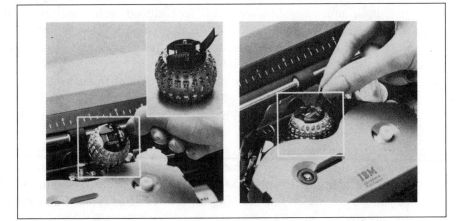

**Figure 5.5. Example of the Relevant Area of a Photo
Indicated by a Contrasting Arrow.** Note how the arrow
in the picture is aligned in the direction that the intended
force is to be applied. (Reprinted from *Volkswagen
Rabbit/Scirocco Service Manual, 1980 and 1981
Gasoline Models Including Pickup Truck* (Cambridge,
MA: Robert Bentley, Inc., 1981), p. 8-66. © Volks-
wagen of America. With permission.)

other disadvantage—realism—cannot be so easily overcome. If you need
a cutaway view, use a drawing. However, when you do use photos, as far
as possible use views that the user can actually see. In so doing, you will
be making the most use of the photo's realism.

One other caution is in order here. For a photo to be of any use to the
reader, it must be shot well and reproduced well. Nothing is more
irritating than trying to make out detail on a photo that is muddy or too
dark. A photo in a manual must have the proper exposure—both as
originally shot and as printed—and it must have adequate contrast.
How well a photo shows fine detail is dependent in part on how it was
originally shot (focus, depth of field, and so on), but it also depends on
the screen used to print it. The coarser the screen, the more detail is
lost.

Photos can be scanned for computer storage, but the images generally
use up a lot of computer memory. If you produce manuals on a smaller
PC- or Macintosh-based desktop publishing system, you may want to
rely more heavily on line drawings or use traditional paste-up techniques
for photos.

Line Drawings

Drawings, or *line art,* are another common choice for manual illustrations. They are easily reproduced on the printed page and require no special processing, unlike photos.

Drawings are popular with manual writers for a number of reasons. You can show exactly what you want to show without having to deal with a clutter of extraneous parts. You can use a drawing to show normally hidden parts (as in a cutaway) or to show assembly sequences (as in an exploded view) (see Figures 5.6 and 5.7). Line drawings can easily be created and stored on a computer.

The chief disadvantage of line drawings is that they are more abstract and require more sophistication on the part of the reader than do photos. Some types of line art—electrical schematics or engineering diagrams,

Figure 5.6. Cutaway View of a Valve. A cutaway view allows the reader to see the construction of an object by presenting it as if a portion were literally cut away, revealing the layers of composition. The cutaway view is particularly useful to show the interior of something that is not normally disassembled. (Drawing by Teresa Sprecher.)

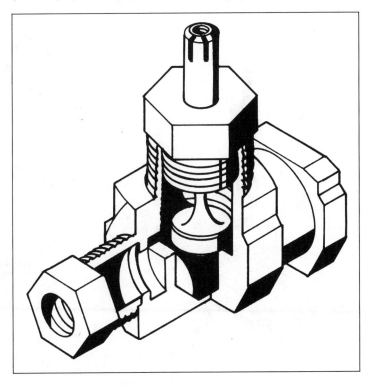

Figure 5.7. Exploded Diagram of a Valve. An exploded diagram allows the reader to see how the parts of an assembly fit together. It is most often used in conjunction with instructions for assembly or disassembly procedures. (Drawing by Teresa Sprecher.)

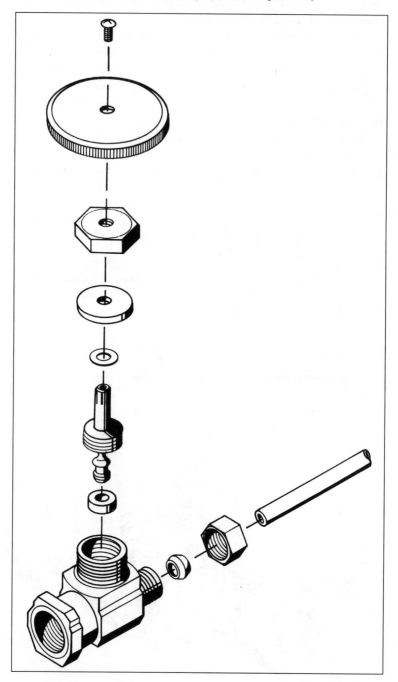

for example—require special training to interpret. The writer must always be aware of his or her audience and make choices accordingly. For example, it is usually a much better idea to provide a perspective drawing of a product to the new owner than to include an engineering diagram (see Figures 5.8 and 5.9.)

Unless your audience is technically sophisticated, leave the technical drawings on the drafting table or at the computer-aided design (CAD) workstation. Even line drawings of perspective views may be difficult for some readers because of the lack of shadows, textures, and details, and the use of drawings to highlight a specific part without background clutter may become a disadvantage if the user cannot then recognize the part in context.

Charts and Tables

Charts and tables are nonpictorial visuals. They are, in fact, a substitute for text. Charts and tables can, if well designed, convey certain kinds of information much more effectively than prose. For example, whenever you have numerical information to present, it will be much easier for the reader to see and keep track of it if you put it in a table rather than a

Figure 5.8. Perspective Line Drawing of a Valve. This drawing is an abstraction—it is not *really* the valve—but it is quite recognizable to an untrained eye. (Drawing by Teresa Sprecher.)

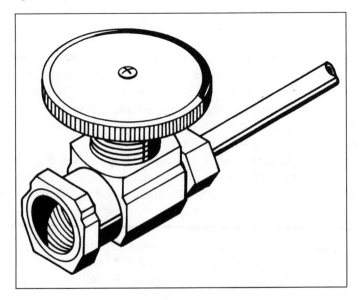

Figure 5.9. Engineering Drawing of a Valve. This drawing is much more abstract and requires special training to interpret. (Drawing by Teresa Sprecher.)

paragraph. Numbers buried in paragraphs of prose are easy to miss and hard to relate to one another. Similarly, just as putting instructions in a series of numbered steps makes it easy for the reader to follow them, putting information that the reader must relate to other information (such as operating conditions and frequency of lubrication) in a table or chart makes it easy to see the connections.

Charts and tables, while not pictorial, *are* visual. Visuals can be seen as a whole, although it may take closer inspection to see the details. Prose, on the other hand, is essentially linear—you can skim it, but you cannot see it all at once. Visuals make it very easy for the reader to locate the information he or she needs. You can glance at a table and find the column of interest right away, whereas it might take several minutes of skimming text to find the paragraph you need.

Charts and tables, then, are commonly used when a particular reader will need only part of the information presented. Typical applications of charts and tables include the following:

- Specifications
- Model comparisons
- Trouble-shooting guides
- Maintenance intervals

Charts. Increasingly, manuals are including *flow charts* adapted from computer programming to present testing procedures in an easy-to-follow format. For example, Figure 5.10 shows a diagnostic testing procedure in flow chart form rather than the more familiar list of steps. Figure 5.11 shows a flow chart for trouble-shooting a labeling machine instead of the more traditional table as shown in Figure 5.12.

Many different kinds of charts may be useful in manuals. With all of them, a few basic rules apply:

1. Keep them as simple and focused as possible.
2. Make it clear to the reader how to use the chart, especially if it is a nontraditional way to present familiar categories of information.
3. Include sufficient white space so the reader can easily locate the needed information.

Tables. Manuals use tables frequently because tables can show a great deal of quantitative information in a very compact form. For example, Figure 5.13 shows a table that gives cooling system capacities for 11 different engines — all in a space less than 3×4 in. Imagine the confusion that would result if you tried to present all that information in prose!

For tables to work effectively, they must be designed in accord with these guidelines:

1. Arrange the headings and data in a rational order.
2. Display items to be directly compared vertically. Most people find it much harder to compare information arranged horizontally (see Figure 5.14).
3. Include units of measurement in headings.
4. Align columns of numbers on the decimal point.
5. Use lines to divide columns and rows only if confusion is likely without them. Lines add to clutter. If possible, use white space as a divider instead (see Figure 5.15).

Figure 5.10. Diagnostic Testing Procedure Shown as a Flow Chart. The flow-chart format makes it very easy to see the sequence of tests. (Reprinted from *Powertrain Diagnostic Procedures A-604 Ultradrive Automatic Transaxle (1989)*, No. 81–699–9009 (Center Line, MI: Chrysler Motors Corporation, 1988), p. 45. With permission.)

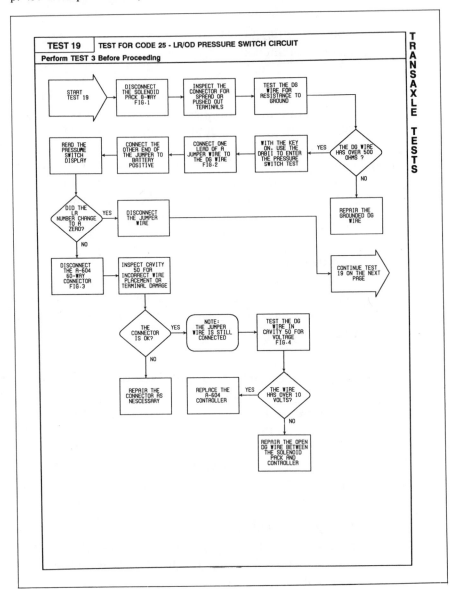

Figure 5.11. Trouble-Shooting Chart in Flow-Chart Format. The flow-chart format gives a clear view of the appropriate sequence of steps. (Reprinted from *Basics of Rotary Labeling, Second Edition* (Franklin, WI: Krones, Inc., 1986), p. 39. With permission.)

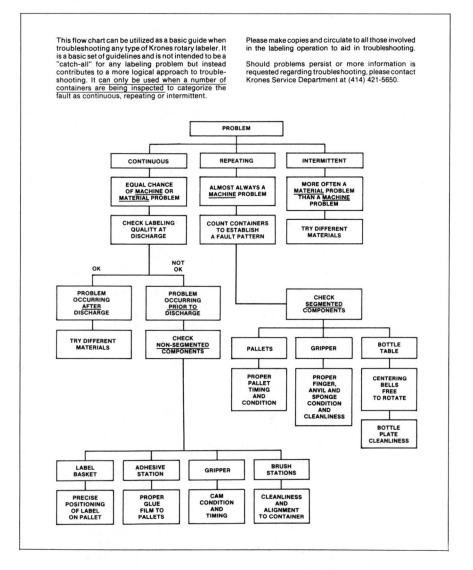

Figure 5.12. Trouble-Shooting Chart in Traditional Format. The traditional format works well when there is no need for a sequential approach to trouble-shooting. (Reprinted from Model 323-1 Tortilla Folder, Technical Service Manual (Davenport, IA: Kartridg Pak Company, 1990), p. 34. With permission.)

PROBLEM	POSSIBLE CAUSE	POSSIBLE SOLUTION
No filling.	Filler nozzle is not primed.	Run about 4 tortillas through to prime.
	Filler metering cylinder is not primed.	Follow instructions under "Set-Up".
	Filler hopper interlock is not being made.	Make certain hopper is properly located & anti-rotation pin is fully engaged.
	Completion of stroke sensor has not been activated.	Use screw driver blade or wrench to push on pilot button.
	Air is off to machine or filler.	Check & turn on.
First fold & fill keep cycling.	Water droplets on electric eye lens.	Clean & dry lens.
	Residue buildup on 1st conveyor belt.	Clean off belt.
	Sensitivity set too high on 1st electric eye.	Adjust so eye only detects passing of tortillas.
Loosely folded burritos.	Tortillas are stale.	Use fresher tortillas. Best results are obtained when tortillas are only a few hours old.
	Back Stop on 2nd fold platen is too far back.	Adjust position to lessen surface area on platen.
	Too small amount of filling.	Adjust filler per pages 16 & 17.

Figure 5.13. Table Showing Data in a Compact Form. The use of a table permits a great deal of information to be packed into a small space. (Reprinted from *Pierce Chassis Operation Manual, Part Number 90–5000* (Appleton, WI: Pierce Manufacturing Co., Inc., 190), p. 23. With permission.)

COOLING SYSTEM CAPACITY			
ENGINE MODEL	ENGINE	RADIATOR	TOTAL CAPACITY
8.2 Dash	12.5	19.0	33.5 quarts
6V-92TA Dash	24.5	26.0	53.0 quarts
6-71T Arrow	22.0	28.0	52.0 quarts
8V-71N Arrow	31.0	28.0	61.0 quarts
8V-71TA Arrow	32.0	28.0	62.0 quarts
6V-92TA Arrow	24.5	28.0	55.0 quarts
8V-92TA Arrow	29.0	28.0	59.0 quarts
NT 855	22.0	28.0	52.0 quarts
L10	13.5	28.0	44.0 quarts
3208 N&T	25.0	19.0	46.0 quarts
3406 N&T	24.0	28.0	54.0 quarts
NOTE: TOTAL CAPACITY INCLUDES APPROXIMATELY 2 QUARTS OF COOLANT NEEDED TO FILL HOSES.			

Designing Visuals

The first task in designing an effective visual, as in designing an effective piece of writing, is to figure out the purpose of the visual. What exactly do you want to show? Do you want an overall shot of the product simply to allow the reader to recognize the model you are going to discuss? Is the visual suppose to show the location of major subsystems or a single adjustment screw? Is the purpose to show how something is assembled or what it looks like in place? You must define as precisely as possible the job a particular illustration is to do. If you have trouble specifying one overriding purpose, maybe you are asking one picture to do the work of several. Consider using more than one visual if you have more than one purpose; otherwise, you may end up with an illustration that, in trying to do everything, does nothing effectively.

Once you have determined the purpose of the visual, you can begin to make the design choices that will ensure the visual works—because you know what is important. The essence of good visual design may be summed up in the following rule: *Make the important things stand out.* For example, depending on what you consider to be important,

Figure 5.14. Two Versions of a Table Showing the Importance of Vertical Comparison.

PARDEE SPINNING RODS

MODEL	LENGTH (FEET)	LINE WT (LB-TEST)	LURE WT (OZ)	ACTION	PRICE
UL-500	4.5	2-4	1/16-1/8	LIGHT	$32.50
UL-600	5	2-4	1/16-1/8	MEDIUM	$38.00
WB-650	6	6-8	1/8-1/4	MEDIUM	$43.00
WB-800	6.5	6-8	1/8-1/4	STIFF	$47.50
BN-850	6	8-12	1/4-3/8	MEDIUM	$52.75
MS-900	6	12-16	1/4-1/2	MEDIUM	$63.00
MS-1000	6.5	12-20	3/8-5/8	STIFF	$72.00

(VERSION A)

PARDEE SPINNING RODS

	UL-500	UL-600	WB-650	WB-800	BN-850	MS-900	MS-1000
LENGTH (FEET)	4.5	5	6	6.5	6	6	6.5
LINE WEIGHT (LB-TEST)	2-4	2-4	6-8	6-8	8-12	12-16	12-20
LURE WT. (OZ)	1/16-1/8	1/16-1/8	1/8-1/4	1/8-1/4	1/4-3/8	1/4-1/2	3/8-5/8
ACTION	LIGHT	MEDIUM	MEDIUM	STIFF	MEDIUM	MEDIUM	STIFF
PRICE	$32.50	$38.00	$43.00	$47.50	$52.75	63.00	$72.00

(VERSION B)

you may decide to use a drawing rather than a photograph or one view rather than another. If you are designing a table or a chart, you can tell how to set it up. As far as time and budget permit, make these decisions for each illustration individually. Don't use an old photo from another manual just because it is handy. In the long run, an illustration or table designed with a specific purpose in mind will better serve the user's needs and the company's interests. The following general principles apply to any visual presentation—drawing, photo, table, or chart.

Figure 5.15. Two Versions of Two Tables — With and Without Lines Dividing the Columns. Note that the use of lines is essential when the white space between columns is irregular — but a distraction when the white space forms a sufficient visual barrier.

LINE CAPACITY OF SPOOL (IN YARDS)		
Lb. Test	Large	Small
2	---	300
4	400	250
6	350	200
8	300	150
10	250	100
12	200	---
15	150	---
18	100	---

LINE CAPACITY OF SPOOL (IN YARDS)		
Lb. Test	Large	Small
2	---	300
4	400	250
6	350	200
8	300	150
10	250	100
12	200	---
15	150	---
18	100	---

PARTS LIST (PARTIAL)			
Part Name	Order No.	Part No.	Price
Axle	305	81-214	$2.25
Baffle plate	6342-R	56	1.20
Click spring	11	322-D	.30
Drive gear	452-T	9120	4.50
Pivot	6783-DE	3	.75
Rotating head	66	81-615-A	11.45
Spool	43598-0S2	27	3.75
Transfer gear	452 (453)	16 (17)	.75 (.85)
Trip lever	340972	81-005	1.25

PARTS LIST (PARTIAL)			
Part Name	Order No.	Part No.	Price
Axle	305	81-214	$2.25
Baffle plate	6342-R	56	1.20
Click spring	11	322-D	.30
Drive gear	452-T	9120	4.50
Pivot	6783-DE	3	.75
Rotating head	66	81-615-A	11.45
Spool	43598-0S2	27	3.75
Transfer gear	452 (453)	16 (17)	.75 (.85)
Trip lever	340972	81-005	1.25

Make Them Big

The user should easily be able to read and interpret a visual at normal reading distance. Remember that an operator manual may be used in less-than-ideal conditions. Depending on the product, the user may be reading the manual in a basement or a dimly lit barn. In any case, the user will probably be glancing from the manual to the product and back again. This makes it important for the user to be able to find his or her place in the visual rapidly — a maneuver that is much easier if the visual is large. For many visuals, it is also crucial that the reader be able to identify parts. If a drawing is too small, a machine screw may look just like a small bolt.

In planning the layout, try to make spaces to fit the visuals rather than the other way around. Photo-reduction is a wonderful thing, but it can ruin a photo or drawing if used indiscriminately. In a photo, depending on the coarseness of the halftone screen used to reproduce it, excess reduction can obliterate important detail. In a drawing, too much reduction can cause letters to fill in and closely spaced lines to run together. Be particularly careful of reducing large assembly drawings or schematics. Typically a line weight that works fine in the original becomes much too light when it is reduced to page size. If the drawing was done on a CAD workstation, you may be able to alter it on the computer, but many times, the better procedure is to redesign the drawing specifically for the manual. Another useful technique is to use close-ups or insets to show detail. Figure 5.16 shows an example of this technique. Be sure that if you use a close-up of a part, the reader can tell how the part relates to the rest of the product. A nice close-up of the power steering fluid filler cap is not much use to the car owner if he can't find it when he looks under the hood!

Make It Simple

Without question, the most common fault of illustrations in operator manuals is that they are cluttered. A user will find it difficult to focus on (or even figure out) what is important in a visual overloaded with information. Illustrations and charts must be edited just like prose: figure out what the purpose of the visual is and then include only what is necessary to fulfill that purpose. You should not include everything you know in a visual—any more than you would in a paragraph. Simplify the visual presentation so that only essential items are included in detail and nonessential items are either absent or merely suggested (see Figure 5.17).

If, despite your efforts to keep it simple, you still seem to have a complex illustration, consider splitting it in two—break the presentation down by systems or show one overview and one or more close-ups. This is an especially important technique to use in parts catalogs which too often show the entire product in a single incomprehensible assembly drawing.

This process of simplification is terribly difficult—there is a great temptation to include more than you should—but you will find that knowing the purpose of the visual will help enormously. If you have a clear idea of what you want the drawing, photo, or chart to accomplish, you can use that idea as a filter to screen out peripheral information.

For example, the block diagram shown in Figure 5.18 is perfectly

Figure 5.16. Use of a Close-Up to Show Detail. Note how the use of a close-up permits much more detail to be shown than would be possible otherwise. (Reprinted from *LCB 13150/16150 Posi-Stop Liquid Cooled Brake Maintenance and Service Manual Supplement* (Statesville, NC: Clark Components International, 1989), p. 4. With permission.)

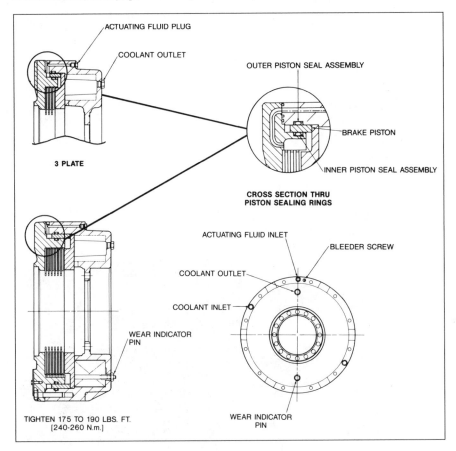

adequate for explaining the theory of operation of the transaxle in this service training manual. The full schematic (Figure 5.19) is included at the end for reference.

Finally, use plenty of white space around and within the visual. This alone will help to reduce visual clutter and make the visual more inviting to use.

Make Them Clear

Be sure that each drawing, photo, table, and chart has a title that tells what it shows as well as a figure number or table number. For example,

"Figure 3, Location of Idle Adjustment Screw" is much better than just "Figure 3". Label the parts of the illustration adequately—and as close as possible to the parts referred to so the reader's eye jumps back and forth as little as possible. If you can, put the labels in words on the drawing or photo itself rather than use callouts (numbers or letters listed in a key elsewhere). Sometimes, however, direct labeling is not possible or desir-

Figure 5.17. Example of an Illustration with the Background De-emphasized. Note how fading out the background permits the reader to concentrate on the relevant portion of the diagram. (Reprinted from *Hydrapower Integral Power Steering Gear TAS40, TAS55, TAS65 Service Manual* (Lafayette, IN: TRW Ross Gear Division, 1989). With permission.)

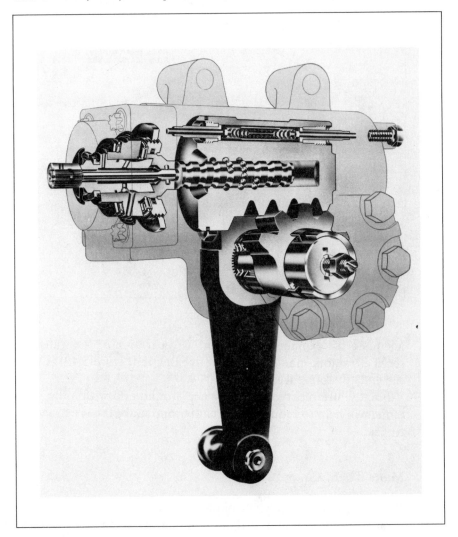

Figure 5.18. Block Diagram to Indicate Electrical Components. At this point in the manual, a block diagram is sufficient. Using a block diagram permits the reader to concentrate on the important parts and avoid being distracted by detail. (Reprinted from *A-604 Ultradrive Electronic Automatic Transaxle Student Reference Book (1988)*, No. 81-699-9011 (Center Line, MI: Chrysler Motors Corporation, 1988), p. 19. With permission.)

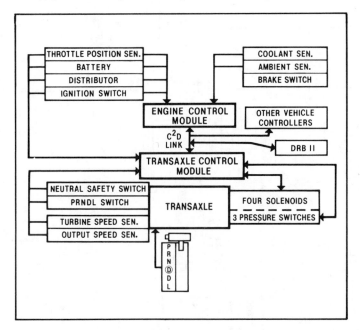

able, as, for example, if labels would clutter the illustration or if the manual is to be prepared in more than one language. If at all possible, orient labels to read horizontally—or at least use not more than two or three orientations.

Use the principles of good visuals design to help focus the reader's attention and to avoid confusion:

- The eye moves with lines, not across them. Use line direction to lead your reader's eye to the central focus of the visual.
- Bigger or more detailed objects will seem more important than smaller or less detailed ones.
- Similar shapes (in a block diagram, for example) will suggest similar function.

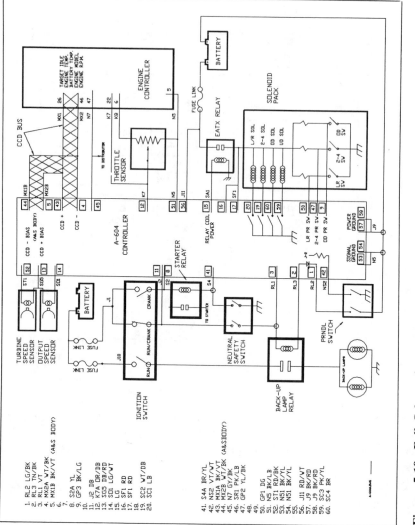

Figure 5.19. Full Schematic to Aid in Servicing Electrical Systems. Here the schematic is necessary for completeness. (Reprinted from *A-604 Ultradrive Electronic Automatic Transaxle Student Reference Book (1988)*, No. 81-699-9011 (Center Line, MI: Chrysler Motors Corporation, 1988), p. 118. With permission.)

Integrating Visuals with Text

Always plan your visuals at the same time as you plan your text. While you work on the outline of your text, be thinking about where to include visuals. Make sketches as you develop a rough draft. Many companies use the "storyboard" in planning manuals. Storyboards are a step between outline and draft: for each small section of the outline— corresponding to no more than a page or two in the finished manual— the writer creates a plan. The plan contains a short prose summary of the material to be covered in that section and a sketch of an accompanying exhibit. That way, the visuals are integrated to the manual from the start, helping to ensure that text and illustrations will balance and support each other.

As in the prose of the text, try to follow a general-to-specific order for illustrations: start with an overview, and then move to subsystems and close-ups of individual parts.

Always refer to your visuals in the text, and always place the visual as soon as possible after the first reference to it.

New Uses of Visual Media

In recent years, visuals have gained prominence as a primary means to communicate information rather than as mere supplements to text. Two new developments in technical documentation illustrate this trend: the development of visual-based manuals and the use of videotapes in connection with or in place of traditional manuals.

Visual-Based Manuals

Some manuals are built entirely around visuals. John Deere and Company is credited with starting this trend with its development of the "Illustruction" format (Figure 5.20). In this format, each separate instruction is tied to a visual—normally a photograph. Manuals are built in terms of these text-visual blocks. Since John Deere introduced the format, many other industries have adapted it for their manuals (Figure 5.21).

Visual-based manuals have several advantages: they look very accessible and work well with poor readers; they require little work to translate into other languages; and they make modular organization easy. They do have drawbacks, however. The rigid one-instruction-per-visual format leaves little room for explanation of complex procedures or functions.

Figure 5.20. Sample Page from a Deere and Company Manual Showing the "Illustruction Method." Note how each photo is keyed to a corresponding block of text, which is kept quite short. The small numbers in the lower right-hand corners of the blocks are the computer "addresses" of the blocks. Thus, relevant segments may easily be used in different manuals. (Reprinted from John Deere *Operator Manual* (OM RW 15455) Issue A1 (Waterloo, IA: Waterloo Tractor Works), p. 31. With permission.)

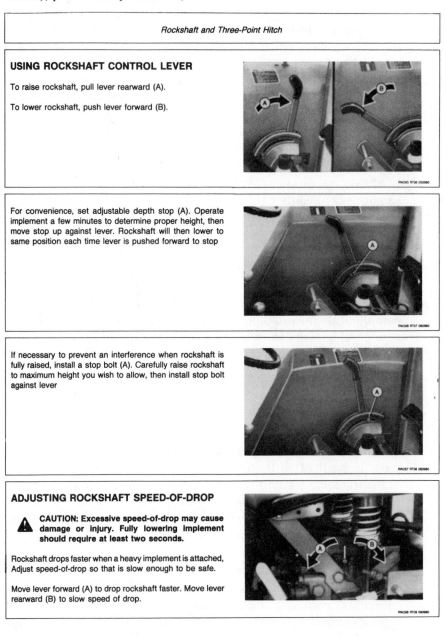

Rockshaft and Three-Point Hitch

USING ROCKSHAFT CONTROL LEVER

To raise rockshaft, pull lever rearward (A).

To lower rockshaft, push lever forward (B).

For convenience, set adjustable depth stop (A). Operate implement a few minutes to determine proper height, then move stop up against lever. Rockshaft will then lower to same position each time lever is pushed forward to stop

If necessary to prevent an interference when rockshaft is fully raised, install a stop bolt (A). Carefully raise rockshaft to maximum height you wish to allow, then install stop bolt against lever

ADJUSTING ROCKSHAFT SPEED-OF-DROP

⚠ CAUTION: Excessive speed-of-drop may cause damage or injury. Fully lowering implement should require at least two seconds.

Rockshaft drops faster when a heavy implement is attached, Adjust speed-of-drop so that is slow enough to be safe.

Move lever forward (A) to drop rockshaft faster. Move lever rearward (B) to slow speed of drop.

Figure 5.21. Variation on the "Illustruction"-Type Format. Here the idea of coupling a single instruction with a visual is used, but with a somewhat different layout. (Reprinted from *34000 Powershift Transmission Maintenance and Service Manual, R Model—12 Speed* (Statesville, NC: Clark Components International, Inc., 1989), p. 10. With permission.)

Figure 53
Remove high idler gear bearing retainer plate bolts and retainer plate.

Figure 56
Tap input shaft and inner bearing from cover.

Figure 54
Remove idler gear and outer taper bearing.

Figure 57
Remove idler shaft front cover plug.

Figure 55
Remove forward clutch front bearing. Remove idler gear inner taper bearing. Remove idler shaft lock plate screw and lock plate.

Figure 58
Tap idler shaft from front cover.

Because the visuals carry the communication burden, they must be extremely well designed and well produced. Finally, unless the company can do most of the work of photography and layout in-house, visual-based manuals can be expensive to produce.

Despite these drawbacks, the visual-based format is gaining increasing use, probably for two chief reasons: the marketplace is increasingly global, requiring manuals in many different languages, and concerns about product safety and liability prevention make it desirable to rely less on words—which can be interpreted in many ways—and more on pictures, which can be (though not necessarily) less ambiguous.

Videotapes

Another technique that is gaining popularity is the use of videotapes as substitutes or supplements for written manuals. The obvious advantage of videotapes is that the viewer can actually watch a procedure being performed by an expert. It can be the next best thing to the "expert-in-residence" that we discussed at the beginning of this chapter. A viewer of a videotape does not need to be literate or skilled in interpreting technical drawings.

On the other hand, the videotape requires that the user have the appropriate equipment to view it. Further, unlike a paper manual, the user cannot easily "leaf through" a tape. For these reasons, we find that videotapes are most useful as training tools for service technicians or as supplements for one-time use, as in showing how to set up a product for operation.

To work well, videos must be carefully planned and produced—and these are not simple tasks. Developing a good videotape means putting a great deal of time into designing it and finding skilled technicians to shoot and edit the tape. Videotapes produced by amateurs look like home movies and almost always present information much too fast. If your company is thinking of using videos, it is well worth the investment to contract with professionals unless you already have the skills in-house. Do-it-yourself videos are usually much less effective than a well-written manual.

Summary

As you can see, choosing and designing good visuals for an operator manual follows much the same process as the writing itself: you must determine your purpose and audience and then choose the visuals in terms of those. You must select and organize information that is presented visually, as you must for information that is presented verbally. You must develop sketches or visual drafts, and, finally, you must "edit" your tables and illustrations according to the rules for visual clarity. The process is the same because the function is the same — words and pictures combine to help the reader learn how the product works and how to operate and maintain it properly; only the language is different.

Review Checklist and Exercise

Checklist

The following is a list of questions to ask yourself about the visuals you have included in your manual.

- ☐ Have I planned what visuals to include from the user's point of view?
- ☐ Have I tried to look at my product with a fresh eye?
- ☐ Have I defined carefully the purpose of each visual?
- ☐ Have I chosen the type of visual to use (drawing, photo, chart or table) based on that purpose?
- ☐ Have I designed my visuals to be big, simple, and clear?
- ☐ Have I taken complex visuals and tried to break them down into simpler illustrations?
- ☐ Have I used adequate white space?
- ☐ Have I carefully labeled the parts of the visuals?
- ☐ Have I tied the visuals clearly to my text?

Exercise

The sample instructions that follow might appear in an operator manual for a (very old) sewing machine. For those of you not familiar with sewing machines, here is a brief description. The balance wheel is a large flywheel mounted on the right side of the machine. As the needle (left) goes up and down, the balance wheel turns. A spool of thread is mounted on top of the machine, and the thread goes through various devices to maintain proper tension and finally goes through the needle itself. Underneath the base of the machine is a small metal spool of thread called the bobbin, which rests in a removable metal case. Thread from this bobbin is fed up through a hole in the bed of the machine and interacts with the needle-controlled thread to produce stitching.

the manufacturing process meets legal requirements (see also Chapter 1, Planning).

Product liability litigation may be based on one of three legal concepts: *negligence, breach of warranty,* and *strict liability in tort. Negligence* means that the manufacturer did not exercise reasonable care in the manufacture or marketing of a product resulting in an unreasonable risk to the user. *Breach of warranty* means that the product did not do what the manufacturer said it would – but note that the warranty can be either express or implied. An implied warranty may be present in instructions for the product's use, even if the matter in question is not included in the written (or express) warranty. *Strict liability in tort* means that if a product is defective in manufacture or marketing, the manufacturer may be liable for damages even if the manufacturer was *not* negligent, was unaware of the defect, and made no claims about the product's performance. Strict liability in tort concentrates on the product, not on the care with which the manufacturer operated. Whatever legal concept is used, four things must be true for a product liability suit to be successful.[1]

- The product must have a defect.
- The defect must be present when the product leaves the control of the manufacturer.
- Injury or damage must be incurred.
- The injury or damage must have been caused by the defect.

All four must be true. For example, an electric drill that overheats dangerously with normal use is obviously defective, but there are no grounds for a liability suit unless someone is injured. If someone inadvertently knocks the defective drill off a workbench so that it falls and breaks his toe, there is still no basis for a claim because the injury was not caused by the defect. A product defect may be a design defect, a manufacturing defect, a packaging defect, a marketing defect, etc. In this chapter we will deal only with defects involving instructions and warnings. These fall into three categories:

- Failure to warn at all of risks or hazards present in the product
- Failure to warn adequately of such risks or hazards
- Failure to provide appropriate and adequate instructions for use of the product

Let's look more closely at this duty to warn.

Who Must Be Warned

The manufacturer has a duty to warn anyone who might reasonably come into contact with the product. Thus, the manufacturer may be held liable for injury to someone other than the person who actually bought the product — e.g., an employee of a company that bought an industrial cleaning product or the child of a consumer who bought an electric coffee grinder. Especially if the likelihood of injury is great or the potential injury is serious, the manufacturer may be required to include a warning directly on the product itself. The situation for the technical writer is further complicated by the fact that the potential users of a product may comprise a very diverse group (see Chapter 2, Analyzing the Manual User). You must consider the possible range in terms of age, sex, expertise, familiarity with product, even literacy. In *Hubbard-Hall Chemical Co. v. Silverman*, the court held that a written warning was not adequate because it failed to provide for illiterate users.[2]

What Must Be Warned About

The manufacturer has a duty to warn potential users of dangers present in the nature of the product in normal use *and* in foreseeable misuse of the product. Thus, a manufacturer of chlorine-containing laundry bleach has a duty to warn of potential skin irritation, since this is a risk inherent in using the product to bleach clothes. However, the manufacturer may also have a duty to warn against mixing the product with ammonia (as someone might do if using the bleach as a household cleaner), which produces deadly chlorine gas, because this is a foreseeable misuse.

The manufacturer has no duty to warn of open and obvious dangers — that a knife cuts, for example. However, you must be sure that the danger is obvious to the user. A danger that is open and obvious to you, who are thoroughly familiar with your company's products, might be unknown to the user. For example, many people are unaware that burning charcoal emits carbon monoxide gas, which can be deadly with inadequate ventilation. Anyone concerned with the manufacture of charcoal briquettes surely knows this, yet every year the news includes reports of people dying while using a charcoal grill in a trailer or closed garage.

Even if the danger is open and obvious, the manufacturer may have a duty to warn if the user may not be aware of the extent or degree of danger. For example, a person using tile adhesive labeled "flammable" probably would not smoke while using the product, but he or she might well not realize the danger posed by pilot lights, especially in remote areas of the home.

Post-Sale Duty to Warn

The manufacturer has a continuing duty to warn of hazards connected with the product, even if the hazard is discovered after the sale of the product. This post–sale duty to warn is most certain when the danger could have been discovered at the time of sale. For example, in *Gillham v. Admiral Corp.*, the manufacturer of a television set began receiving complaints of a fire hazard from a high-voltage transformer in the set very soon after the set was put on the market. By the time of the fire that prompted this suit (1968), the manufacturer already knew of 91 instances in which fires had occurred. Yet the manufacturer had not warned of this hazard. The court decided that this failure to warn left the manufacturer liable — both for compensatory damages and punitive damages.[3]

A stickier situation arises when the danger is discovered long after manufacture and sale of the product. This can be a particularly thorny problem with prescription drugs or other medical products, in which unexpected side effects can take a long time to appear. We have seen a number of such cases in recent years. On the face of it, it seems quite unreasonable to hold a manufacturer liable for dangers that were not known and could not be known at time of sale. Remember, however, that under strict liability in tort, a manufacturer has a positive duty to provide a safe product and therefore also (at least to some courts) has a later duty to warn of dangers unknown at time of sale.

An interesting sort of post-sale discovery of danger is the danger that results from unforeseeable use. You will recall that the manufacturer has no duty to warn of unforeseeable uses — yet once that use has been called to the manufacturer's attention, it is no longer unforeseeable. A good example is *Temple v. Velcro USA, Inc.* The manufacturer learned of hot-air balloonists using Velcro in hot-air balloons. The company had certainly not designed the closure for such purposes and considered them to be inappropriate and dangerous. It promptly mounted a campaign to warn against the practice, enlisting the aid of the Federal Aviation Authority and sending warnings to registered balloonists. The court held in the manufacturer's favor when it was sued for the wrongful death of a balloonist *who had read and disregarded the warning.*[4]

Obviously, the practicality of warning present-day users of products manufactured and sold years ago varies with the product. With a drug, the manufacturer can warn physicians who might prescribe it. With an industrial lathe that has been resold several times, or a household blender, the problem is much more difficult. Sometimes the best that a manufacturer can do is make a good faith effort to warn users of newly discovered hazards or, especially when the danger is great, to offer to repair or replace the hazardous component.

Older products are of particular concern to manufacturers because safe use of older products often depends heavily on the operator's being aware of hazards and taking steps to avoid them. Manufacturers have no guarantee that anyone but the original owner of a product will receive the instruction manual. In addition, safety equipment on older products was often easily removed or bypassed. For example, a recently manufactured drill press might be designed so that the power cannot be turned on unless a movable safety shield is in place. An older drill press might include the safety shield, but make its use optional. Courts have awarded damages to persons injured by machinery from which safety shields have been removed because removal of safety equipment was considered to be a foreseeable misuse.[5] These cases point to the need for warnings placed on the product itself, as well as in the instruction manual — particularly if your company's product is likely to be resold or used by persons other than the original purchaser.

What Is an Adequate Warning?

To be considered adequate, a hazard warning must do four things:

- Identify the gravity of the risk
- Describe the nature of the risk
- Tell the user how to avoid the risk
- Be clearly communicated to the person exposed to the risk

All four of these elements must be present. For example, in a case involving asbestos fiber hazards, the court found a warning inadequate because it did not state the severity of the risk and the nature of the hazard.[6] The warning read as follows:

> This product contains asbestos fiber. Inhalation of asbestos in excessive quantities over long periods of time may be harmful. If dust is created when this product is handled, avoid breathing the dust. If adequate ventilation control is not possible, wear respirators approved by the U.S. Bureau of Mines for pneumoconiosis-producing dusts.

Actually, this particular warning could serve as a model of how *not* to write a warning. We will return to it later as an annotated example.

Strategies for Warning and Instructing

Warning Labels on the Product

When you are developing warning labels to be placed on the product itself rather than in the manual, you should try to design labels that follow these guidelines:

- Make your warnings consistent.
- Never mix general instructions with warnings.
- Follow existing guidelines.
- Make sure warnings meet all applicable standards.
- Place the warning near the hazard.
- Make sure the warning is readable.
- Make sure the label is durable.

Let's look at each of these guidelines in turn.

Make Your Warnings Consistent. This principle applies not only to warnings, but, as we have seen, to general instructions as well (see Chapter 3, Organization and Writing Strategies). By following a consistent format, you set up a pattern of expectations in your readers — they expect to find the same sort of information in the same places from warning to warning. As long as you remain consistent, you remove from your readers the burden of figuring out the *type* of message you are giving them, allowing them to concentrate their attention on the message itself.

Never Mix General Instructions with a Warning. A warning may be overlooked or the seriousness of its message may be diluted if it is combined with general instructions. This is generally more of a problem in instruction manuals, but sometimes labels contain a mixture of warning information and other information that would normally also be included on a product label. For example, a label might read "WARNING: steam under pressure can cause severe burns. Release pressure before removing radiator cap. Keep radiator filled to appropriate level with 50/50 mixture of recommended coolant and water." The last sentence of that warning is related to the hazard, since an underfilled radiator will build up more pressure than a filled one, but it does not relate to the immediate hazard of escaping steam. The shorter the better — as long as all necessary information is given about the nature and severity of the risk and the means to avoid it.

Follow Existing Guidelines. Two particularly helpful sets of guidelines are put out by the FMC Corporation and by Westinghouse. These may be obtained from those companies.[7] The guidelines suggest designing warning labels consisting of three parts:

- A signal word and color to convey the severity of the hazard
- A symbol or pictogram showing the nature and consequences of the hazard
- Words to describe how to avoid the hazard

Figure 6.1 shows some sample warning labels.

These guidelines suggest that a single word be used consistently to convey a particular level of risk:

- DANGER (red): the hazard or unsafe practice *will* result in severe injury or death.
- WARNING (orange): the hazard or unsafe practice *could* result in severe injury or death.
- CAUTION (yellow): the hazard or unsafe practice could result in *minor* injury or property damage.

These guidelines conform to the American National Standards Institute draft standard ANSI Z535.4, "American National Standards for Product Safety Signs and Labels," issued in January 1989.

Other instructions or cautionary statements pertaining to optimum use (e.g., "after 25 hours of service, change the oil in your lawn mower's crankcase") should use other signal words, such as *notice*.

The symbol or pictogram should depict the nature of the hazard and its consequences, preferably by showing both the machine part and the body part. FMC has developed a series of pictograms to cover various common hazards (see Figure 6.2). The use of symbols or pictograms in addition to a verbal message is particularly important if the product in question may be used by persons who are illiterate (including children) or who do not speak the language in which the warning is written.

The final element of the warning label, the verbal message telling how to avoid the hazard, must be clearly communicated to the reader. The wording should be simple, direct, and active. Do not be afraid to word a warning strongly and specifically. Do *not* say: "May result in bodily harm;" *do* say: "can amputate fingers."

Figure 6.1. Sample Warning Labels. (Reprinted from *Product Safety Label Handbook* (Pittsburgh, PA: Westinghouse Electric Corporation, 1981). © 1981. With permission.)

Figure 6.2. Sample Pictograms. (Reprinted from *Product Safety Sign and Label System*, 3rd ed. (Santa Clara, CA: FMC Corporation, 1980), p. 7–2. With permission.)

At this point, let's return to the warning about asbestos fibers and analyze it in terms of these verbal guidelines.

This product contains asbestos fiber. Inhalation of asbestos in excessive quantities (*How much is that?*) over long periods of time (*How long?*) may (*Possible, not probable?*) be harmful (*How harmful?*). If dust is created when this product is handled, avoid breathing the dust (*How?*). If adequate ventilation control is not possible (*What's adequate?*) . . .

Clearly, the reader of this message is given very little real information. Some manufacturers fear that writing forceful, specific warnings will make their products seem unreasonably dangerous. In general, potential buyers are glad to have specific information, and the courts have held numerous times that a vague warning is an inadequate warning. A successful product liability suit can cost a company a good deal more than the loss of a couple of sales to persons "scared off" by well-written warnings.

Make Sure Warnings Meet Applicable Standards. If standards for warning labels exist that apply to your particular industry, failure to observe them will automatically make your warning inadequate in the eyes of the court. Standards vary widely from industry to industry and often do not offer specific, practical help in designing warning labels. Nevertheless, you must become familiar with whatever standards govern your area.

Place the Warning Near the Hazard. It is not enough to design a good warning label if the user of the product gets hurt anyway because he or she failed to notice the warning. For example, a Kentucky court awarded a $266,000 settlement to a farm worker injured when his hand was pulled into a corn picker.[8] A 3 × 5-in. decal on the side of the picker warned to disengage the power before cleaning the rollers — but there was no warning decal near the rollers. Neither the farm worker nor the owner (who was running the tractor that powered the picker) saw the decal.

Be Sure the Warning is Readable. Consider how far away from the product the user will be before you choose your type size — both to increase the likelihood that the user will read the warning and to protect his or her safety. You don't want the user to have to come dangerously close to a hazard just to be able to read the warning! Think about what kinds of lighting conditions are likely when the product is used: will the light be dim (as in a poorly lit basement or barn) or glaringly bright (as in a sunny corn field)? Choose colors and label material accordingly. What is the viewing angle? Will the user be able to look directly at the warning where you plan to place it? If the angle is too severe, the words may not be readable. Try to foresee all the likely conditions of use for your product — and then make your design decisions.

Make Sure the Label is Durable. A carefully designed warning label is useless if it dissolves in the rain, is rubbed off in a few weeks of use, or is totally obliterated by a few smudges of oil or dirt. In choosing the materials you use for your labels, you must consider the material to which the label will be attached, the length of time it will have to last, and the kind of treatment it will receive. Be sure that the materials match — i.e., that the base material and the adhesive are compatible and of comparable durability (there is no point, for example, to a long-lived base material and a short-lived adhesive). Be sure that the ink used to print the message will set well with the base material. Make these decisions early, and consult with suppliers for specifications on various materials.

Instructions and Warnings in Manuals

Instructions for proper use of a product must be clear, readable, and understandable by readers. Follow the guidelines provided in the rest of this book for creating good instructions. Clear and complete writing is the best way to avoid inadequate instructions.

Warnings included in the text of an instruction manual must follow many of the same guidelines as labels on the product itself. They must be

- Internally consistent
- Consistent with the labels on the product itself
- Consistent with applicable standards

You may have many more warnings in the text than labels on the product, but be sure that every DANGER-level warning in the text (hazard or unsafe practice that will result in severe injury or death) has a corresponding label on the product. Be careful that you do not overuse the DANGER designation—too many of them may dilute the impact. If you find yourself writing an inordinate number of DANGER warnings, it may indicate that the product is unreasonably dangerous and should be redesigned.

Never mix safety warnings with ordinary instructions, and never bury warnings in the text in such a way that they might be missed. Make sure warnings included in the text stand out and are easily readable.

At the same time, avoid grouping all the safety warnings together on one page at the beginning or end of the manual. First of all, if they are all on one page, it is all too easy for the reader to skip that page and miss the warnings. Second, if the warnings are grouped together, they will not be in front of the reader when he or she is working through a procedure. We recommend that you put warnings in the text wherever they are relevant—and that you repeat a warning if it applies to more than one situation. Make sure your readers see the warning before they act.

If your company's legal department insists on having a page of warnings at the beginning of the manual, go ahead . . . but put them in the text as well, wherever they are needed. If you do have a "safety page" at the front of the manual, be sure to organize it so that warnings about related hazards are grouped and labeled with a heading.

Standardized Warnings and Labels

"Why not make all labeling and safety instructions uniform? Then there would be no mistakes." Many products resist neat categorizing or are composed of components not all of which are suitable for standardized warnings and labels. You have only to read the safety instructions for three or four different brands of baby toys, riding mowers, or power tools to see how different manufacturers vary in their choices for warnings and labels. Even so, some standards exist, mostly for narrowly defined and very specific products, such as hazardous industrial chemicals.

Benefits of Standardized Warnings

Standardized warnings have a number of advantages:

- Warnings and word choice have been subjected to committee review.
- Warnings represent a composite of the judgments of safety specialists in specific industries.
- Warnings are backed up by combined experience of many companies who make the product.
- The accumulation of case law and liability experience helps to identify trouble areas.

Those who favor this committee approach to creating standardized warnings also believe that repetition of the same warning symbol or words is itself a plus. They argue that confrontation with the same warning again and again serves to reinforce the message and to promote instantaneous recognition.

Limitations of Standardized Warnings

Checklists or "cookbook" warnings also have drawbacks largely because the cookbook warning is only as good as the cooks who produced it. Even committee-designed warnings can reflect blind spots or errors in judgment, and a bad standardized warning can gain widespread use simply because it is standardized. Further, uncritical reliance on pre-designed warnings can be deceptively easy and sometimes even dangerous. If you begin to "plug in" standard warnings without paying attention to the special characteristics of your product, safety problems may begin to slip by. Your product may only seem to be the same as others, when in reality you need to think creatively about its hazards.

Esthetic Trade-Offs

Some industries resist the color, size, or wording of standardized labels on esthetic grounds. For example, kitchen ware and appliances whose color and design are part of the sales appeal may suffer esthetically from the warning label. Manufacturers may choose to emboss glass or use raised lettering on plastic rather than stick a blazing red label on an avocado green product. Trade-offs between esthetics and standardized warnings are common where sales of a product depend heavily on eye appeal.

Safety Cartoons

A difficult issue is the use of safety cartoons. The main attraction is that they are eye-catching. Most people will look at a cartoon in preference to reading a written warning. However, in our view, cartoons have a number of serious drawbacks, and we do not recommend their use. The shortcomings of safety cartoons include the following:

- *They may dilute the warnings in the text*—especially if the warnings incorporate pictograms, as we recommend. You cannot include a cartoon for every warning, but including cartoons for only some may imply that the other warnings are trivial.
- *They may seem to treat a heavy subject lightly* and thus undermine their own purpose. The cartoon format may remind product users of Saturday morning TV cartoons, in which terrible things happened to the cartoon characters without rendering them any permanent harm.
- *They are very hard to design well.* To effectively focus attention on a hazard, a cartoon must be extremely simple and uncluttered and must make the hazard itself clear and obvious.
- *They may imply that you are talking down to readers*, thus alienating them and making it less likely that they will read (and heed) other warnings or instructions for safe practice.

Given all these shortcomings, and given the availability of effective symbols and pictograms (which are also eye-catching), we recommend that safety cartoons be avoided.

Other Safeguards for the Manufacturer

Two other areas exist which the manufacturer may use to further protect against product liability suits: front matter in the manual and documentation that the user has received the manual. Both of these areas are legally important in proving that the user of the product got the proper instructions for safe use of the product.

Front Matter

Every instruction manual should be dated and should tell exactly what model product it covers and what previous books, if any, it replaces. The manufacturer is therefore protected from the claim that the user did not know that it was the wrong book for the product. Second, the front matter should include appropriate disclaimers. These may or may not protect the manufacturer in a given situation, but they cannot hurt. The disclaimers should state that:

- No warranties are contained in the manual other than . . . (State whatever warranty is appropriate for the product.)
- The instruction book does not alter any agreement for division of responsibilities worked out between the manufacturer and the dealer—so that the manufacturer is not held liable for something the dealer should have done.
- The information in the manual is not all-inclusive and cannot cover all unique situations.

Documentation

The manual should include, in a conspicuous place (such as inside the front cover), some documentation that the buyer received the appropriate instruction manual with the purchase of the product. This may take the form of a postcard filled out and signed at time of sale and sent to the manufacturer or some other form, but it should be included—especially in any manual for a product that is inherently hazardous. The manufacturer then has proof that it fulfilled its duty to provide instructions and warnings at least to the immediate purchaser. This does not, of course, guarantee that the information will be passed along to other users.

Review Checklist

Here is a list of questions to ask yourself about the safety warnings and messages that will go with your product. If you can answer yes to all the questions, you have probably done an effective job of helping to protect your product's users and your company.

- ☐ Have I identified and warned about all the hazards connected with my product?
- ☐ Have I looked at my product through the eyes of a first-time user?
- ☐ Have I anticipated foreseeable misuses of my product?
- ☐ Have I included warning labels on the product itself for severe hazards and for hazards connected with a product likely to be resold or used by someone without access to the instruction manual?
- ☐ Do my warnings include all four elements of an adequate warning?
- ☐ Have I separated warnings from general instructions?
- ☐ Have I followed existing guidelines?
- ☐ Do my warnings meet applicable standards?
- ☐ Have I placed my warning labels near the hazards they warn against?
- ☐ Will the warning labels be readable during normal use of the product?
- ☐ Are the labels durable? Will they last as long as the product does?
- ☐ Have I included appropriate warnings in the text of the instruction manual?
- ☐ Are these consistent with any labels on the product itself?
- ☐ Do they stand out from the rest of the instructions?
- ☐ Have I included appropriate information in the manual about the model(s) it covers, the book(s) it replaces, or the date it becomes effective?
- ☐ Have I included appropriate disclaimers?
- ☐ Is there a means to document that the buyer received the manual?

Summary

The rapid increase in product liability litigation in the last decade and the case-law approach to defining product defectiveness have without question put a tremendous responsibility on manufacturers to evaluate product safety efforts. This situation has led to the development of much more sophisticated and standardized practices in writing instructions and safety warnings and thus raised the quality of these materials in general. The role of technical writers has become more important and more respected — and the job has become harder. The guidelines presented in this chapter should help you to design effective warnings and write helpful instructions to ensure that people use your company's product safely.

References

1. Ross, K. "Legal and Practical Considerations for the Creation of Warning Labels and Instruction Books," *J. Products Liability* 4:29–45 (1981); see also "Section 402A," *Restatement of the Law Second, Torts 2d*. Philadelphia: American Law Institute Publishers, 1965.
2. *Hubbard-Hall Chemical Co. v. Silverman*, 340 F 2d 402 (1st Cir. 1965).
3. 523 F.2d 102 (6th Cir. 1975) (applying Ohio law), *cert. denied*, 424 U.S. 913 (1976); discussed in Roddy, N. E. "The Product Manufacturer's Post–Sale Tort Responsibilities: Warnings, Recalls, Repairs," *Product Liability Trends* 14:97–106 (1989).
4. 148 Cal. App. 3d 1090, 196 Cal. Rptr. 531 (1983); discussed in Roddy, N. E. "The Product Manufacturer's Post-Sale Tort Responsibilities: Warnings, Recalls, Repairs," *Product Liability Trends* 14:97–106 (1989).
5. *Craven v. Niagara Machine & Tool Works, Inc.*, 417 N.E. 2d 1165 (Ind. App. 1981).
6. *Borel v. Fibreboard Paper Products Corp.*, 493 F 2d 1076 (5th Cir. 1973); Cert. denied, 419 US 869 (1974); see also Ross, K. "Legal and Practical Considerations," p. 35.
7. *Product Safety Sign and Label System*, 3rd ed. (Santa Clara, CA: FMC Corporation, Central Engineering Laboratories, 1980); *Product Safety Label Handbook* (Trafford, PA: Westinghouse Electric Corporation, Customer Service Section, Westinghouse Printing Division, 1981).
8. *Moore v. New Idea Farm Equipment Co.* 78-C1-069 (Lee County, Ky. Cir. 1981).

7

Service Manuals

Overview

Although service manuals employ many of the same principles and techniques as operator manuals, the two forms differ. A service manual has a much more specialized audience and purpose, and this difference is reflected in text and design. This chapter will discuss these differences, as they are reflected in content, style, visuals, and mechanics. Certainly, the principles of good verbal and visual design outlined elsewhere in this book will still apply: instructions should be presented in parallel form, visuals should be clear and easy to read, and so on. However, the application of these principles will differ, and this chapter will present detailed guidelines for putting the principles to work in the specialized context of the service manual.

How Service Manuals Differ from Operator Manuals

Purpose

The purpose of a service manual is very different from that of an operator manual. An operator manual is intended primarily to give clear instructions for a product's use and care. It introduces a new user to a product and explains what the product is for and how to make it work. An operator manual may give simple maintenance procedures, such as how to clean the cabinet of a TV set or how to change the oil in a lawn mower, but such instructions usually cover only the most basic operations. For anything complicated, the user is usually referred to "an authorized service representative." A service manual, in contrast, is what the authorized service representative uses. The purpose of a service manual is to explain in detail the repair and maintenance of a product – to explain, for example, how to clean and adjust a faulty tuner in a TV or how to overhaul the engine of a lawn mower. Normally, a service manual will assume that the reader is familiar with the product, knows how to operate it properly, and needs specialized information only.

In addition, the service manual often serves as the "textbook" for training technicians. In factory training programs and in technical schools, students learn by doing. A student in a transmissions class at a technical college, for example, will learn how to repair auto transmissions by working on one chosen from a particular make and model of car. The service manual for the car will be the primary resource for the student learning how to do a particular procedure.

Audience

The audience for a service manual is also very different from that for an operator manual. As we have already discussed, the audience for an operator manual may be anyone from a professional user who is technically sophisticated to a member of the general public who perhaps has never even seen the product before, much less used it. The audience for a service manual, in contrast, is almost always technically sophisticated and very familiar with the product. This audience will most often consist of professional repair and service technicians, but may also include knowledgeable amateurs – the do-it-yourself auto mechanic, for example.

Contents

What a Service Manual Does Not Contain

Unlike an operator manual, a service manual will usually not include any introduction to the product. It will not cover what the product looks like, what its major components are, what it is used for, or what its capabilities are. Since the reader is assumed to be familiar with the product, this kind of information is not considered necessary. If the reader of the service manual is of the "knowledgeable amateur" category, he or she may need to rely somewhat on the operator manual for this kind of background information.

What a Service Manual Does Contain

Naturally, the precise content of the service manual will vary with the product, but certain categories of information will appear in all service manuals. These include the following:

- Specifications for the product, including capacities for lubricants, cooling agents, etc., and recommendations for lubricants and cleaning agents to be used
- Technical background on the function and operation of the product or its systems
- Routine maintenance procedures and recommendations for service intervals
- A trouble-shooting guide
- Repair procedures
- Model change information
- A parts catalog (this may be a separate publication)

This chapter will discuss each of these categories in detail. First, however, we would like to make some general remarks about how these different kinds of information are treated in a service manual. The information in a service manual will naturally be far more complete than in an operator manual. More complete explanations will be given, and more complex procedures will be described. These procedures may require specialized tools. It is common practice for factory–produced service manuals to refer to tools by factory numbers instead of generic names (e.g., "VW 558" instead of "flywheel holding fixture"). The writer of a service manual must be aware that *not all* users of the service manual will be authorized, factory-trained technicians. It might appear that using a specialized terminology for specialized tools would discourage amateurs; however, it is more likely that the amateur will simply find some other way to do the procedure (e.g., using a screwdriver rather than a snap-ring pliers). Example 7.1 shows how this issue has been addressed in one auto service manual. If a specialized tool is *necessary* for a given procedure for safety reasons or to avoid damage to the product, note this fact and refer to the tool by name as well as number.

Safety information in general is important in a service manual. The more complex procedures covered in the service manual are often also more inherently hazardous. In addition, the technician may grow careless through familiarity with the hazards and not take proper precau-

Example 7.1. Accommodating Nonprofessional Users. (Reprinted from *Volkswagen Rabbit/Scirocco Service Manual, 1980 and 1981 Gasoline Models Including Pickup Truck 1981* (Cambridge, MA: Robert Bentley, Inc., 1981), p. 5–61. © Volkswagen of America. With permission.)

> 3. Install the flywheel holding fixture on the pressure plate assembly, as shown in Figure 10–3. Alternatively, you can use coat hanger wire to bind a bolt hole in the engine block's transaxle mounting flange.[1]

tions, or the technician, having performed similar procedures many times, may not read instructions closely and may miss vital safety information unless it is prominent. All of the guidelines given in Chapter 6 on safety warnings also apply to service manuals.

Specifications

Service manuals will contain detailed information on specifications and tolerances. Operator manuals will contain some of this same information, but only what is necessary for routine maintenance by the owner. For example, the operator manual for a cassette recorder would include specifications for power input from various sources (batteries, house current, car battery), with specifications for appropriate adapters, or a car owner's manual might include specifications for spark plug gap, on the assumption that the owner might do his or her own tune-ups. However, a service manual would be much more detailed. Example 7.2 shows the great detail with which a typewriter service manual describes adjustments to the on-off switch.

Example 7.2. Detailed Specifications in Text. (Reprinted from *Single Element Typewriter, Model 200, Service Manual* (Torrance, CA: Silver-Reed America), p. 5. With permission.)

> 1. The Motor Pulley Ring should be positioned at a distance of 2.5–3.0 mm from the end of the Motor Shaft.
> 2. The clearance between the Motor Pulley and the Motor Pulley Washer should be 0.1–0.2 mm.[2]

Information about specifications and tolerances will be found in three contexts in a service manual: in the text of technical background and procedures sections, in tables, and in visuals. (Refer to Chapters 3 and 5 and to the section in this chapter covering visuals for information about the designing of drawings and photos and the proper setup of tables and charts.)

When you include specification information in technical background and procedure sections, you must be especially careful to avoid:

- Letting the numbers get lost in the paragraphs of text
- Letting the numbers obscure the flow of your description of how an assembly works or how a mechanism should be adjusted.

This requires constant attention to how numbers relate to the rest of the text. If you have one or two numbers in a long paragraph of explanatory text, the reader can all too easily skim over them and miss what may be vital information. On the other hand, a paragraph loaded with numbers is terribly hard to read. Generally, more than four or five exact numbers in a paragraph of text is too many. The reader simply cannot keep the numbers straight and often loses the line of thought conveyed in the text.

You may use a number of techniques to solve these problems. To make an occasional numeral stand out in a sea of words, print it in boldface. A manual with different typefaces may be a little more expensive to produce, but is well worth it if the extra expense ensures that the material is used. As an alternative, the numbers may be separated by white space from the surrounding text. If the number occurs in a procedure description, try to put the exact number in a step of its own rather than including it with other adjustments. Example 7.3 shows how this technique can clarify the text.

Example 7.3. Separating Steps in Directions.

Original

1. To adjust idle, carefully turn knurled adjustment screw no more than one-quarter turn at a time, until idle speed of 500 rpm is reached.

Improved

1. To adjust idle, turn knurled adjustment screw in or out. Do not turn more than one turn at a time.
2. Adjust idle to 500 rpm.

If the text contains many exact numbers, it may be better to put them in a separate table or chart. A set of exact numbers can be much more easily assimilated in chart form than in paragraph form (see Chapter 5, Visuals, for more information on tables and charts.) Example 7.4 shows two versions of a description of how to adjust automobile wiper arms, one of which uses a chart to separate verbal from numerical information.

The last technique for making numerical information visible is to include tolerance and specifications in visuals accompanying the text. In designing your visuals, you must be careful that numerical information does not clutter up the drawing. A good drawing can easily be ruined with too many labels and excessive specification information, especially in a service manual. Since the audience is generally more technically sophisticated than the audience for an operator manual, the writer may be tempted to use unedited engineering drawings for visuals. Although such drawings contain a wealth of specification information, they are

Example 7.4. Using a Chart to Separate Numerical Information. (Version 2 reprinted from *Dodge Dart, Coronet and Charger Service Manual, 1967* (Detroit, MI: Dodge Division, Chrysler Motors Corporation, 1967), p. 8-89. With permission.)

Version 1

With the force applied, the clearance between the tip of the wiper blade and the windshield lower moulding should be $1/2$ to 2 in. on the right and $1/4$ to 2 in. on the left for the Dart, and $1/2$ to $2^{1/2}$ in. on the right and $1/4$ to $2^{1/4}$ in. on the left for the Coronet and Charger.

Version 2 (Actual)[3]

With the force applied, the clearance between the tip of the wiper blade and windshield lower moulding should be as follows:

Models	Clearance in Inches Between Tip of Blade and Windshield Moulding	
	Right	Left
Dart..................	$1/2$–$2^{1/2}$	$1/4$–2
Coronet and Charger	$1/2$–$2^{1/2}$	$1/4$–$2^{1/4}$

For most readers, the second version will be far more usable.

Figure 7.1. Example of a Visual Used to Convey Specifications and Tolerances. (Reprinted from *Single Element Typewriter, Model 200, Service Manual* (Torrance, CA: Silver-Reed America, Inc.), p. 44. With permission.)

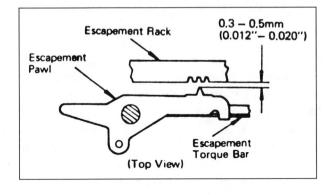

Figure 7.2. Example of a Visual That Conveys Information About Tolerances. Note the effective use of a close-up to avoid cluttering the drawing and to make it more readable. (Reprinted from *Single Element Typewriter, Model 200, Service Manual* (Torrance, CA: Silver-Reed America, Inc.), p. 45. With permission.)

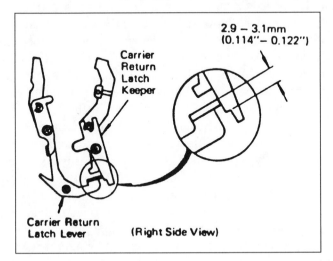

usually too cluttered to be useful to the service technician. It is certainly possible to use visuals well to convey specification information, as Figures 7.1 and 7.2 show. Notice especially the use of the close-up to illustrate a particular portion of the drawing. Be sure, if you use visuals to present specification or tolerance information, that the visual is placed right next to the relevant text, particularly in procedures sections.

Model Change Information

Service manuals, unlike operator manuals, are normally updated periodically. The product user does not need update information, since he or she is unlikely to buy successive versions of the same product. In cases where that does happen—a company, for example, may replace office machines on a regular basis—the manufacturer usually supplies a new operator manual with each product. This approach is impractical, however, for the service technician who repairs and maintains a manufacturer's product line over a period of time. One technician, for example, may service several generations of a particular manufacturer's small gasoline engines—as well as those of other manufacturers. Especially when year-to-year modifications are relatively minor, it is much cheaper and handier for the manufacturer to supply supplements to an existing service manual than to write a whole new manual each time the product changes slightly. The writer of a service manual must keep in mind this need for frequent updates and must design the manual so that updating is easily accomplished.

Similar problems appear when a company's products are predominately custom installations, as may be the case with, for example, packaging equipment or medical equipment. Somehow, the company still has to provide an accurate service manual. Computer technology is making this easier, particularly with a modular manual: one can, with the right system, custom-assemble a manual for each machine. For most companies, this degree of automation is a ways in the future, and more traditional methods are the norm, both for customizing and updating.

Many different techniques are used to update manuals. Some manufacturers supply replacement or supplemental pages to be inserted into the service manual which, of course, requires that the service manual be bound in a loose-leaf binder. Increasingly, manufacturers are using video and computer technology to keep their service technicians informed of changes and new procedures. For example, an auto manufacturer may supply its dealerships with a videotape showing a new service procedure being performed, rather than requiring the dealership to send a mechanic to the factory for training. Or technicians may be provided supplemental information on microfiche. No doubt, as the computer gains even wider use in businesses, new information will be sent out in the form of floppy disks or CD-ROM. Although all these technological improvements in communication are desirable, the fact remains that not all users of the service manual have access to the technology to use them. Even if such technology is available, the initial "high-tech" communication should be followed up with "hard copy"—including supplements to the manual. These supplements are usually in the form of individual pages or separate pamphlets.

We recommend the former—replacement or add-on pages—for two reasons:

1. If new information is bound right into the original service manual, the manufacturer is assured that the technician has the new information. Separate booklets are too easily lost or misfiled.
2. If the new information is bound into the manual at the relevant place, it is more likely to be noticed.

To ensure that the update pages are used, the writer of a service manual can use the following techniques:

- At the beginning of the manual, tell the reader that supplements will be provided from time to time and explain how to use them. Distinguish between add-on pages, which should *follow* existing pages in the book, and replacement pages, which require removal of the old pages.
- Provide a page at the beginning or end of the book on which the technician can record the addition of supplemental material.
- Number pages clearly according to the original manual's pagination and build in a means of distinguishing add-on from replacement pages. (For example, add-on pages might be numbered with the page number of the page they should follow, plus a letter— 26a, 26b, etc.—whereas replacement pages would simply be numbered the same as the pages they are supposed to replace.)

The writer of a service manual should also build in ways to draw the technician's attention to model change information. For example, if add-on or replacement pages are likely to appear in the manual, you may wish to put a reminder at the beginning of each chapter to look for model change information, or you may wish to have update information printed on stock of a different color from the original page stock.

Field Modifications

You should also include supplemental information about probable field modifications. This information tells the service technician of ways in which the owner may have modified the product. Although these modifications are often not approved by the manufacturer, they do take place, and the technician should be made aware of them. If a particular modification appears often, it may be a signal that a design change is needed in the product — particularly if the modification involves the removal of safety equipment. See Chapter 6 for some discussion of this problem. Sales representatives and local dealers are good sources of information for the service manual writer about what modifications may be expected.

Parts Catalog

A service manual may also contain a parts catalog for the product, although this is often a separate publication. If your manual does include a parts catalog, you must be careful that the principles of good visual design apply. Refer to Chapter 5 for details, but, in general, follow these guidelines:

- Be sure the drawings or photos are clear and large enough to see easily.
- Label each part, preferably with name and order number.
- If you use a system of callouts on the visual combined with a separate list of parts, be sure the list is next to the visual so that the user can refer easily from one to the other.
- If the parts catalog is separate from the manual, be sure that any changes in design that alter the parts catalog are reflected as update information in the manual.

Style

All of the organizational and writing strategies described in Chapter 3 apply to service manuals as well as to operator manuals. The use of such techniques as general-to-specific organization, lists, parallel structure, and active voice are just as important to the reader of the service manual. The differences in style between an operator manual and a service manual have to do primarily with level of language and tone, rather than with the basic principles of presenting information.

The language in a service manual will be a good deal more technical than the language in an operator manual. Since anyone reading a service manual has a certain amount of technical expertise, the writer can use a more specialized vocabulary. Don't make things technical just for the sake of making them technical, however; good writing of any sort is as simple as it can be and still convey the necessary information concisely. Remember also that not all readers of the manual will be familiar with the manufacturer's particular name for things. If you use a term that is special to one manufacturer, try also to include the generic name for the item as well.

The pace of a service manual may also be somewhat faster than that of an operator manual. This simply means that you can present information at a faster rate on the page. You may include less background information and more substantive words per sentence. Remember, however, that the emphasis in a service manual is on procedures, which means that the reader will probably be looking back and forth between the manual and the product as he or she performs a procedure. Keep your sentences and paragraphs relatively short, and use formatting devices to make it easy for the technician to find the right place in the manual again after looking away for a moment to do a step in a procedure. Overloaded sentences and paragraphs are just as annoying to a technically sophisticated reader as they are to a first-time operator.

Finally, the tone of a service manual may be less conversational than that of an operator manual. As we have noted, the function of an operator manual is in part to represent the company to its customers and in part to gently introduce the new user to the product. This requires that an operator manual be written in everyday language and that it sound "friendly." The purpose of a service manual is primarily to explain to a professional or knowledgeable amateur technician how to perform various repair and maintenance procedures. Example 7.5 shows the differences in tone. The example contains two excerpts, the first from a typewriter owner's manual and the second from a typewriter service manual, both dealing with the operation of the right margin stop.

To sum up, the stylistic principles involved in writing operator manuals and service manuals are the same; the principles are applied differently, however, as a result of the differences in audience.

Visuals

As with the verbal portion of a service manual, the basic principles for visual design (explained in Chapter 5) apply to both operator manuals and service manuals. Good visual design is perhaps even more important in service manuals because so much of a service manual is devoted to procedures for repair and adjustment. The combination of verbal and visual material must make the procedure perfectly clear. This often means that the balance of material shifts toward the visual: a service manual will tend to have more drawings and photographs than an operator manual.

Although the principles are the same for both kinds of manuals, again their application differs. A service manual will have more technical drawings; exploded diagrams and cutaways rather than perspective drawings; and circuit diagrams rather than block diagrams. You must take great care to ensure that these are large enough to see easily and are not cluttered.

An exploded diagram, for example, can often be made much more comprehensible by dividing it into sections. Figure 7.3 shows how "sectionalizing" an exploded drawing of a transmission permits more complicated portions to appear in close-up. The whole view could have been laid out in one piece and photo-reduced to fit on the page, but the result would have been difficult to read.

As suggested in Chapter 5, when the complexity of the drawing permits, label parts with part name rather than a callout. Since the technician using the manual will already be looking back and forth between the manual and the product, adding another place to look (the key that

Example 7.5. Style Differences between Operator and Service Manuals. (Operator manual reprinted from *IBM Correcting Selectric II Operating Instructions* (White Plains, NY: International Business Machines Corporation, 1973), p. 6. © 1973. With permission. Service manual reprinted from *Single Element Typewriter, Model 200, Service Manual* (Torrance, CA: Silver-Reed America), p. 33. With permission.)

Operator manual:

The right margin stop prevents you from typing past the right margin; however, you can space or tab right through it. To type past the right margin, press MAR REL (margin release) and continue typing.[4]

Service manual:

As the Carrier moves to the right still more after the Bell ringing, the Margin Stop Latch moves up the Margin Stop Right extension allowing the Margin Rack to rotate. Then the Margin Plate attached to the Margin Rack rotates the Linelock Bellcrank through the Margin Link. At that time, the Linelock Bellcrank extension moves the Linelock Keylever downwards causing its extension to insert into the space between the Keyboard Lock Balls. And the Keyboard has been locked to prevent typing past the Margin Stop Right.[5]

Figure 7.3. Example of "Sectionalizing" an Exploded View to Make It Appear Less Cluttered and to Enable the Reader to View Smaller Parts in Close-up. (Reprinted from *Dodge Dart, Coronet and Charger Service Manual*, 1967 (Detroit, MI: Dodge Division, Chrysler Motors Corporation, 1967), p. 21–18. With permission.)

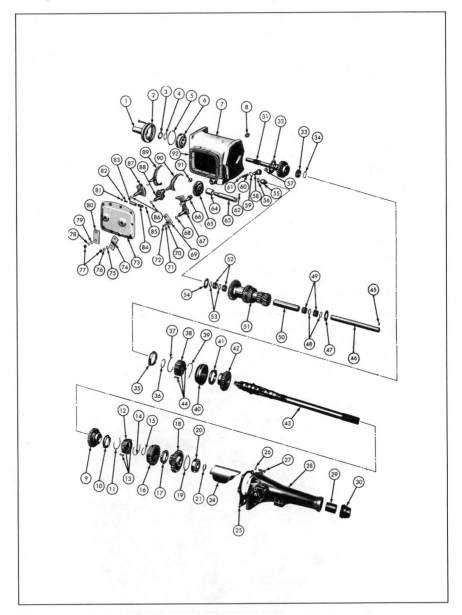

identifies the callout) will only increase the possibility of a mistake. You must also ensure that the lines showing how parts fit together are easily distinguished from lines or arrows leading from labels or callouts. One good way to do this is to use broken lines for the former and solid, heavier lines for the latter.

If you use cutaway drawings, be sure that the reader can easily differentiate the "layers" of the cutaway. Often you can do this by careful shading—but be careful that your shading does not clutter the drawings. At other times, the best choice may be to use color to highlight different levels. Figure 7.4 shows how a photograph (or drawing made to simulate a photograph) can be used as a cutaway. Notice how easy it is to distinguish the parts.

Perhaps the most easily abused form of illustration is the circuit diagram. The writer of a service manual should avoid the temptation to use an unaltered version of the schematic developed with the product. First of all, the original schematic was probably drawn on a large scale or designed on a computer-aided design (CAD) station to be plotted on a large scale. Reducing it to fit onto the manual page would render it unreadable. Second, it may contain more detail than the technician

Figure 7.4. Example of a Photograph Cutaway. Note how easy it is to distinguish the different parts because of the photographic appearance. (Reprinted from *Dodge Dart, Coronet and Charger Service Manual,* 1967 (Detroit, MI: Dodge Division, Chrysler Motors Corporation, 1967), p. 5–37. With permission.)

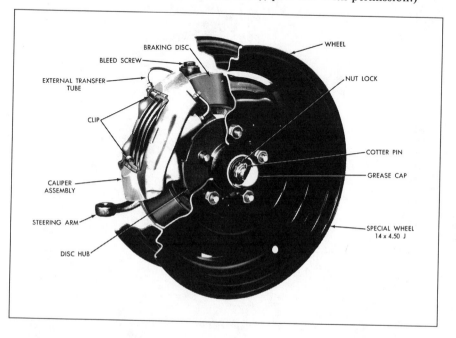

needs, which may lead to unnecessary clutter. Instead, have a schematic drawn or edited for the manual, one that includes only necessary information and is scaled appropriately for the manual page size.

If your product's manual includes a circuit diagram, you must be sure that your readers can interpret the symbols used. For a device that is primarily electronic, this is not usually a problem. Someone without knowledge of electrical circuitry is not likely to use a radio service manual, for instance. However, for a product in which an electrical system is only one component — farm equipment, for example — and for which the service manual's users are likely to be diverse, you may wish to include additional information. For example, the writers of a tractor service manual included the chart shown in Figure 7.5 as explanation for electrical symbols.

Good visual design simply means making sure that your visuals are big enough to be easily seen, are as simple as they can be while still conveying the necessary information, and are clear enough to be easily understood. How these basic principles are put into practice depends on your audience. Since the audience for a service manual is likely to be more knowledgeable than the audience for an operator manual, you can use more technically sophisticated visuals — but you should still make them big, simple, and clear.

Format and Mechanics

Format

Service manuals differ from operator manuals in their large-scale organization and in the relative importance of certain large-scale features. The organization is still determined by "user questions," but the users of service manuals ask different questions. The new owner of a product is concerned with how the product works, how to care for it, and so on. The service technician wants to know how to service or fix the product. The new owner will probably flip through the pages of the operator manual, reading more or less at random. The service technician will look for a specific section that covers the necessary procedure and read that section only, unless specifically directed elsewhere. Thus, a service manual must be organized to help the reader locate the procedure or explanation needed for a particular job and to direct the reader to other relevant sections.

We noted in Chapter 2 that most operator manuals are organized by general categories of information, such as start-up procedures, maintenance, storage, and so on. Generally service manuals are organized by

Figure 7.5. Page of Electrical Symbols Included in a Manual to Help Readers Find Their Way Through Circuit Diagram. An excellent example of writing with the user in mind. (Reprinted from *Service Manual, Series 2 Four-Wheel Drive Tractors, Applicability: 1977 Production* (Winnipeg, Manitoba, Canada: Versatile Farm Equipment Company), p. 6-4. With permission.)

SYMBOL	MEANING	SYMBOL	MEANING
	Wires crossing. No connection.		Mechanically actuated switch: normally closed, held open.
	Wires connected.		Multi-position rotary switch. Connections and positions as tabulated in diagram.
	Ground connection		
	Meter or gauge, as labelled.		Relay, single pole.
	Motor, DC.		
	Lamp, illuminating.		Solenoid and Valve
	Lamp, indicating.		
	Battery: two or more cells.		Resistor, fixed.
	Switch, general.		Resistor, variable.
	Pushbutton, normally open.		Fuse, current rating as labelled.
	Pushbutton, normally closed.		Circuit breaker, current rating as labelled.
	Thermostat switch, closes on rising temperature.		Compressor clutch.
	Pressure sensor, closes on rising pressure.		Speaker.
	Pressure sensor, opens on rising pressure.		

product system: for example, an auto service manual will have chapters on the engine, the transmission, the cooling system, the electrical system, and so on. This kind of organization makes finding the information needed quite simple. If the car has a problem with the cooling system, the technician knows exactly which chapter to read. Often each chapter will have its own table of contents, as in Figure 7.6.

This organizational system works very well when the technician knows the precise location of the problem. However, the systems of a product interact and the cause of the malfunction may not be immediately obvious. Because the service manual divided up by section does not so easily

Figure 7.6. Sample Table of Contents for a Chapter in a Manual. This kind of sectionalizing keeps the main table of contents from becoming too cluttered and is another example of general-to-specific organization. (Reprinted from *Service Manual, Series 2 Four-Wheel Drive Tractors, Applicability: 1977 Production* (Winnipeg, Manitoba, Canada: Versatile Farm Equipment Company), p. 2–1. With permission.)

SECTION 2:   ENGINE SYSTEMS

Table of Contents

1   **INTRODUCTION**

2   **DESCRIPTION AND OPERATION**

2.1   Engine 2- 3
2.2   Engine Mounts .... (Ref. Section 8)
2.3   Fuel System 2- 3
2.4   Cooling System 2- 5
2.5   Air Intake / Exhaust System 2- 8
2.6   Engine Lubrication System 2-10
2.7   Engine Electrical System .... (Ref. Section 6)
2.8   Cold Start System 2-11

3   **TROUBLESHOOTING**

3.1   General 2-13
3.2   Fuel System 2-14
3.3   Cooling System 2-15
3.4   Air Intake / Exhaust System 2-16
3.5   Engine Lubrication System 2-17
3.6   Cold Start System 2-18

4   **INSPECTION / CHECK**

4.1   General 2-19
4.2   Fuel System 2-19
4.3   Cooling System 2-21
4.4   Air Intake / Exhaust System 2-22
4.5   Lubrication System 2-25
4.6   Cold Start System 2-26

5   **MAINTENANCE**

5.1   General 2-28
5.2   Servicing .... (Ref. Section 1)
5.3   Replacement of Fuel Hoses 2-28
5.4   Replacement of Fuel Tanks 2-29
5.5   Replacement of Fuel Gauge Sender 2-29
5.6   Removal and Installation of Surge Tank 2-30
5.7   Removal and Installation of Radiator 2-31
5.8   Replacement of Cooling System Hoses 2-33
5.9   Flushing of Cooling System 2-35
5.10   Removal and Installation of Air Cleaner 2-35

show the overlap of systems, the writer must take care to include a comprehensive trouble-shooting section that directs the reader to the appropriate pages on the basis of symptoms rather than identification of the problem. In other words, instead of listing only "carburetor adjustment, . . . p. 3–23" it would also be helpful to include "car hesitates . . . see carburetor adjustment, . . . p. 3–23."

For the same reason, the writer of a service manual must include a comprehensive index. If possible, include cross referencing within the index, e.g., "carburetor, adjustment (see also fuel filter)." The idea is simply to make the manual useful to the technician by making it as easy as possible for him or her to find the section needed.

The problem of how to organize the information in a service manual becomes particularly difficult when the manufacturer decides to combine the service manual and the owner's manual in one book. Although one book is cheaper to produce than two and ensures that everyone has the same information, we do not recommend this practice. As this chapter has shown, the service manual has a very different audience and purpose than the operator manual, and this difference should be reflected in content, style, and format. To combine both kinds of manuals in one book makes it nearly impossible to maintain the appropriate distinctions. Furthermore, including the service manual with the operator manual may encourage some owners to perform procedures they should not perform because they aren't skilled enough or don't have the appropriate tools. We believe that it is much better for the manufacturer to keep operator manuals and service manuals separate: if the owner *is* a skilled amateur, he or she can always write the company and order the service manual.

Mechanics

Many of the same considerations apply to the mechanics of a service manual as apply to operator manuals, but some differences should be noted. First, a service manual will probably get much harder use than an operator manual. Once an owner has learned how to use and care for the product, the operator manual will probably lie untouched unless a problem occurs. The service manual, in contrast, will be used day in and day out. Even if the technician is thoroughly familiar with a particular procedure, he or she will still need to check information on tolerances and specifications. The cover must stand up to this hard use; usually plastic or (at least) coated stock is required for a service manual cover.

We have discussed the importance of making the manual easy to update. For this purpose, we recommend some kind of loose-leaf binding, e.g., ring binding, a clamp-type binding, or a combination of the two. If you use a binding, be sure that the pages are heavy enough or reinforced so that they will not tear out under hard use. In general, the pages of a service manual must be heavier and more soil resistant than those of an operator manual simply because of the harder use expected.

Finally, the service manual is usually bigger than an operator manual: a standard page size is $8^{1}/_{2} \times 11$ in. Since a service manual is usually not used out of the shop, it need not be as portable, and the larger page size makes the drawings easier to read.

Summary

We have seen in this chapter that although the same basic principles apply to the writing of both operator manuals and service manuals, the application of those principles differs because the two kinds of manuals have different audiences and different purposes. Although both manuals may be written about the same product, they will differ in content, style, and organization. The service manual is written for an audience that is more technically sophisticated and is interested primarily in procedures for service and repair. Therefore, the manual will contain information about the technical background of a product system and about procedures for repair. Since the audience is more technically sophisticated, more technical language and a less conversational style may be used. Because of its specialized purpose, a service manual will be organized to assist the technician in finding the exact repair procedure needed. All the differences notwithstanding, the design procedure for both types of manuals is the same: clearly define the audience and purpose and then arrange and write the material to reflect that definition.

References

1. *Volkswagen Rabbit/Scirocco Service Manual, 1980 and 1981 Gasoline Models Including Pickup Truck 1981* (Cambridge, MA: Robert Bentley, 1981), p. 61.
2. *Single Element Typewriter, Model 200, Service Manual* (Torrance, CA: Silver-Reed America, Inc.), p. 5.
3. *Dodge Dart, Coronet and Charger Service Manual, 1967* (Detroit, MI: Dodge Division, Chrysler Motors Corporation, 1967), p. 8-89.
4. *IBM Correcting Selectric II Operating Instructions* (White Plains, NY: International Business Machines Corporation, 1973), p. 6.
5. *Single Element Typewriter, Model 200, Service Manual* (Torrance, CA: Silver-Reed America), p. 33.

8

Manuals for International Markets

Overview

In the last two decades, international trade has grown in volume and complexity. If you have attended national conferences and trade fairs, you are doubtless aware that "Think International" has become a familiar slogan and that the economic links between nations are steadily increasing. As a consequence, even small and intermediate-size companies whose markets have historically been confined to the U.S. now find themselves selling computers in Africa, rice planters in Southeast Asia, and trucks in China.

When a company decides to market its products outside the U.S., its manual, along with its service and marketing publications, may have to be produced in languages other than English. Because English is the predominant language of international trade, competent English-speaking representatives or translators will usually be on hand at the initial negotiation and contract stages. However, when products actually begin to be sold and used in other countries, written materials in the native languages become a necessity. Those companies already involved in international trade expect to produce their manuals in some or all of the following languages:

Afrikaans	French	Japanese
Arabic	German	Norwegian
Danish	Greek	Portuguese
Dutch	Hebrew	Serbo-Croatian
English	Hungarian	Spanish
Farsi	Indonesian	Swedish
Finnish	Italian	Turkish

In this chapter we will discuss some of the special problems with manuals in translation and suggest ways to make the translating job easier and more cost-effective.

International Manuals: The User Spectrum Expands

User analysis is always a challenge because levels of technical competence and literacy, gender differences, and age of the users all have to be considered. International markets compound the challenge. Writers have to think about users in Third World countries, about products designed in temperate North America and sold in the tropics or the desert, and about service and parts replacement in countries where the majority of people may never have owned a car or a telephone.

If international marketing is a new venture for your company, some general observations about the probable characteristics of users may be of help. Think of users as distributed along a spectrum from industrialized nation to developing nation. Given the rapidity of political and economic realignments now going on, some countries will move back and forth along the spectrum, and change will be rapid. Increases in trade within the Pacific Rim (and with Japan, in particular), the new status of Hong Kong in the 1990's, and the formation of the European Economic Community each precipitate change and establish new rules and laws for doing business internationally.

Characteristics of the Industrial Nations

The industrialized nations include North America, Europe, Russia, Japan, and anomalous southern "pockets" such as Brazil, Australia, and New Zealand. The developing nations include most of South and Central America, Africa, and Southeast Asia. (Trade with China is a special case since China only recently entered the world trade market and has not clearly identified or aligned itself in the world economic order.)

The GNP and per-capita income of industrialized countries are higher, technology is more advanced, and consumer demands for American products are more specialized. Products most in demand are those domestically expensive or preferred because the American technology of the product is considered superior. Manual writers can thus make the following assumptions about manual users in industrialized nations:

- Literacy levels will be high.
- Users will probably be familiar with standard technologies.

- Service and repair support systems will be established and available.

Characteristics of the Developing Nations

The GNP and per-capita income of developing countries will be low, the technologies will be scattered and unreliable, and the demand for imported consumer goods will be more diverse — everything from clothing, cosmetics, and baby food to appliances, automotive products, plumbing supplies, and medical equipment. Manual writers can thus make the following assumptions about manual users:

- Literacy levels may be low, and illiteracy may be more prevalent.
- Users may be unfamiliar with technologies — even the simple technologies.
- Service and repair support systems may be spotty, primitive, or nonexistent.

Problems in Producing Translated Manuals

The chief problem of producing good translated manuals is their expense. Translated manuals will be the most expensive per copy of all your publications. Ideally, if your manual needs translation into one or several languages, you should try to create an English original manual that needs a minimum of changes. Companies already involved in multiple language manual production identify these as key problems:

1. Identifying competent translators (human, machine, or combination of human and machine) is difficult. Translators sometimes make promises about accuracy or costs and then fail to deliver a satisfactory translation.
2. Translated manuals are short runs, and fewer copies of small-batch runs mean higher costs per copy.
3. Labeling on visuals must be redone.
4. Nomenclature for certain parts and tools varies from language to language.
5. Service and repair systems vary widely in quality and comprehensiveness from country to country.
6. Parts identification and replacement become more difficult as the supply line lengthens (the more remote the country and the poorer its infrastructure of roads and railway and air travel, the worse the problem).

7. Cultural differences may affect manual use. For example, some countries have culturally entrenched prohibitions against performing certain mechanical and maintenance tasks. Rural areas, in particular, may have scant understanding or familiarity with machines and their uses.

Let's take a look at these problems in more depth.

Translation

Manuals meant for distribution in industrialized countries are obviously easier to prepare because users are more likely to be literate and familiar with technologies. A further advantage is that the majority of English words have their origins in Romance and Germanic languages, the languages used widely in many "Westernized" countries. French, Spanish, Italian, and German are often sufficient for European markets. A number of companies report that Japan is increasingly insisting upon manuals produced in Japanese before the product is accepted. In fact, the "sales" factor is a powerful lever, and the presence or absence of the translated manuals can make a difference in whether the product sells or is approved for import by foreign competitors.

Manuals produced in the more "exotic" languages are harder to deal with. For instance, Russian, Arabic, Greek, Japanese, and Chinese all have their own alphabets or characters. An Arabic manual must be formatted to read from back to front, beginning on the last page, and the lines of text are read from right to left. Most Japanese manual text is arranged horizontally, as in English, but some technical Japanese texts are read vertically. International manuals also require that measurements and specifications be expressed in metric terms, whose conventions are different from English measure. Other conventions also vary from country to country. For example, the punctuation of numbers:

U.S. 1,000,000.025 vs. **European** 1.000.000,025

If you are marketing products in Canada, remember that Canada is bilingual. Canadian law requires that consumer publications (i.e., those expected to be used by the general public), such as operator manuals, be printed in both French and English. English suffices for the more technical publications such as service manuals, to be used by technicians only.

If you are marketing products in Mexico, labeling should of course, be in Spanish. Bilingual labeling is desirable in the U.S. for a number of products, especially those used in agriculture (farm machinery, pesticides, herbicides, fertilizers) where the labor force may be Spanish-speaking. Your labels and warnings on such toxic or dangerous products should be in Spanish as well as English. Remember, too, that "border Spanish" and French-Canadian have many differences from the Castilian Spanish and the Continental French of Europe.

Sometimes companies are able to identify particular ethnic groups as closely associated with a particular trade or industry. For example, one American manufacturer of a paint sprayer that, when misused, could be dangerous, also recognized that a large percentage of its customers were Greek-speaking contract painters. The company now prints some of its instructions and danger warnings in Greek as well as English.

Choosing Translators

Natural language is rich, slippery, and laden with nuance. In the passage from one language to another, the meanings of words are sometimes skewed or miss the mark entirely. For example, an English manual which read, "Secure the 5/8-in. bolt" was translated into German as "Put the 5/8-in. bolt behind bars." The word "secure" was completely misunderstood. In another manual, "hydraulic ram" became "water goat."

Sometimes mistranslation merely produces howlers, like the computer translation of the "The spirit is willing, but the flesh is weak," which the computer translated as "The drinks were abundant, but the meat was rotten." However, mistranslation becomes serious business especially when safety and precision are at issue. Dangerous procedures or processes requiring great precision need to be explained accurately, with no slippage in meaning.

Human Translators

Throughout the U.S., many firms and individuals offer translation services. Finding translators for European markets is relatively easy, but competent translation into the "exotic" languages may take some hard searching. Avenues for finding translators are language and engineering departments in colleges and universities, the yellow pages in telephone books of metropolitan areas, large companies whose products are similar to yours and who have been selling "international" for a long time, and the good old-fashioned "grapevine."

Many companies have found that their own dealers and international representatives can serve as translators or can identify competent translators within the country. The use of company dealers and service representatives as translators is especially valuable because they know the products as well as the language. As one European put it, "I'm not much good at a social event, but I do know tractor English."

Within the U.S., a great deal of manual translation is a kind of cottage industry. Small firms may be, in reality, single individuals who have created a network of consultants specializing in various languages; many of them work at home. The quality of translation varies a great deal and is, of course, dependent upon individual translator skills as well as upon the complexity of the product for which you need the manual. As you look for individuals or firms to help you, be sure to ask if they have experience with manuals.

There is no fail-safe method for assuring competent and accurate translation, but those with experience in the field offer the following advice:

- Use native speakers in the "home country" of the language, if their translating skills are coupled with good knowledge of the product.
- Within the U.S., the translator should not have been away from the "home language" for more than 5 years. Over time, translators begin to lose touch with current idiom and expressions in the native language. They begin to substitute "English-isms" and American idiom.
- If possible, the translator should have a good working knowledge of the technology of the product or, at the least, a natural feel for how things work. A translator whose specialty is 17th century poetry may be a whiz with the language, but a poor choice for a manual for a circuit breaker or a road paver.

Be prepared to pay for language skills. Some translators charge by the word and others charge by the page or job, depending on the nature of the original. Recently, for example, translation costs for a mechanical product's service manual of several hundred pages were $65,000, and, of course, most of the costs for translated manuals are in that first copy. If you must rely on a translator who knows nothing about your product, make sure that a dealer or service representative who knows the language checks it for accuracy.

Machine Translation

The computer age has provided translators with a vast array of tools, such as data bases for foreign language alphabets and vocabularies, dictionaries and indexes for technical terms, metric conversion, and graphics and formatting options, to name but a few. Many of the large corporations have developed extensive, industry–specific machine translation systems. A number of them have ongoing foreign language research projects underway, attempting to speed up and regularize the translating job.

Companies that use outside computer translation firms report mixed results. Some of the translation systems have been found to have vocabularies too limited for the product. Others have found that syntactical and grammar problems in the exotic languages could not be overcome. Manual producers with long experience using computer translation say, at this writing, that the promise of 80 to 100% accuracy in computer translation has, in many cases, been unrealistic and that total accuracy is still a long way off. They also report that machine translation works very well, in the following situations:

- Simple text, limited vocabulary, heavily visual
- Tabular information, graphs, charts
- Modular format

However, even with modular format, written text usually "swells" about 20 to 30% in translation from English to Germanic and Romance languages, but "contracts" in translation from English to Japanese.

In brief, machine translation continues to encounter the unruliness of natural language. If computer translation produces an 80% accurate manual for you, you will still have to rely on human translators for that last 20%. It is a 20% job that many professional translators avoid. One translator commented, "Being given a computer translation is like being presented with a house that looks like a house, until you get inside. Then you discover that the plumbing leaks and the wiring short circuits. I'd rather move out, start over, and build the house myself."

Choosing the best machine translator for your product has as much trial and error as does the choice of human translators. You will need to balance the tradeoffs between cost, accuracy, and time, using machine translation just as you use the other tools of your trade.

Visuals

The most important component of manuals in translation is the visual. When visuals are clear and self-explanatory, they also help to diminish whatever errors or mistranslation might creep into the translated verbal text.

All of the suggestions for good photos, drawings, and charts suggested

Figure 8.1. Manual with Stand-Alone Visual Pages and "Half-Size" Format. These pages rely entirely on the visuals and on symbols used widely in international manuals (OK, X, √, and the safety alert). Horizontal format and "half-size" of manual are designed for glove compartment storage in truck cab. (Reprinted from *TS 60686* (Allentown, PA: Mack Trucks, Inc., 1986), pp. 14, 18, and 19. With permission.)

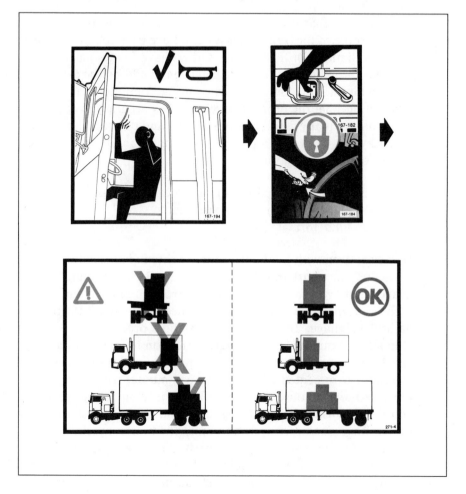

in Chapters 3 and 5 hold true for translated manuals. Visuals should be big, simple, and clear. They should be carefully coordinated with verbal text and planned so that the most important elements are visible (Figure 8.1).

Visuals are particularly crucial when user literacy levels are likely to be low. If you find that, on an average, you are devoting less than half of your manual to visuals, try to add more illustrations. Do this even if, to you, the mechanism or procedure seems perfectly obvious. We recommend, too, that you pay special attention to safety labels and warnings

Figure 8.1. (continued)

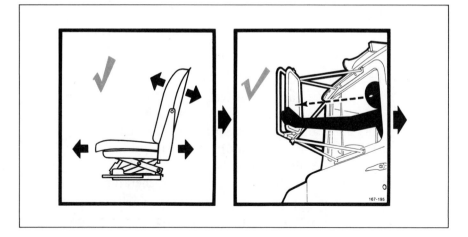

expressed in words alone. The addition of pictograms or drawings to warnings may well make the difference between a user's recognizing a danger and missing it entirely.

You can grasp the importance of visuals by imagining your own reaction to a manual originally written in Chinese and poorly translated into English. Your lifeline to assembly, use, and maintenance of the product will be the photos and drawings.

We suggest in Chapter 5, Visuals, that you use labels directly on the visual whenever possible. Manuals for translation are a notable exception, largely because of the costs. Key numbers and accompanying legend in English can easily be translated by changing the legend only. For visuals in translated manuals:

- Follow suggestions for visual effectiveness in Chapter 5.
- Add more visuals if verbal text exceeds visuals in space allotment.
- Use key numbers and legend to simplify translation of complex visuals.
- If possible, incorporate visuals or pictograms into safety warnings and labels.

Nomenclature

Names for parts, processes, and procedures vary from language to language. Sometimes commonly used English words have no real equivalents. In fact, even "English" and "American differ notably. For example:

English	American
Petrol	Gas
Earth (electrical)	Ground
Flex	Wire
Spanner (generic)	Wrench
Bonnet (auto)	Hood
Boot (auto)	Trunk
Dynamo	Generator

What may be a "shovel" in English may be a "spoon" in another language.

Many "mature" industries, those that for decades have been producing products whose essential features change little from year to year (e.g., plows, automobiles, cameras), have created comprehensive glossaries or dictionaries of routine industry terms. These terms are common parlance, i.e., everyone understands the difference between a knob and a handle and knows what is meant by a socket wrench or a tire iron.

These efforts to regularize language and to create standard vocabularies are often referred to as *Simplified English*, and NCR, Kodak, Ford, Caterpillar, and IBM are some of the companies that have developed these systems.

In looking for consistency, you may be surprised to discover how inconsistent and idiomatic your manuals really are. One editor discovered that manual writers had used seven different terms for a small stop valve on a diesel engine, calling it spacer, washer, shim, stop, intermediate plate, plate, and stop valve. Another writer had used the idiomatic "mule drive" instead of "90° angle belt drive" and another had advised, for fiberglass repair, "Take a piece of rosin the size of a walnut." (In many countries, the walnut example would be meaningless.)

In translated manuals, you should make an effort to standardize the vocabulary for parts and processes and to be consistent in describing routine procedures. For example, do not instruct the user to "oil" the machine at one time and to "lubricate" it the next or to "monitor the needle for pressure variations" the first time and to "check the gauge" the second. (Note that the terms *oil* and *lubricate*, as well as the terms *monitor* and *check*, actually have slightly different meanings in English, but are sometimes used loosely or interchangeably.)

Standardized vocabulary is also important in processes and procedures. The following are some commonly used procedural verbs:

tighten	remove	raise	press
loosen	add	lower	release
fill	check	attach	stop
empty	place	fasten	start
clean	turn	adjust	replace

Make decisions about the procedural words you will need to use, and stick to them. Don't, for example, say "lower the arm" in one place and "allow the arm to drop" in another or "turn the wheel to the right" and then "rotate the wheel clockwise."

Service, Repair, and Parts Replacement

Americans are accustomed to convenient and readily available service and parts replacement. Remember the frustration in the early years of foreign cars in the U.S. when customers complained about waiting weeks for a part? In international trade to industrialized countries, the advent of modern inventory control and computer update has produced vast improvement in the delivery system of service and repair.

Visuals are again of crucial importance in translated service manuals and parts lists. Many sophisticated and advanced systems of manual production have developed parts catalogs and repair manuals that consist almost entirely of visuals and numbers. Parts lists contain only an identifying picture or drawing and an identification number, and service manuals are done almost entirely with photos of the product, photos of the tools needed to assemble and maintain the product, and a limited "universal" vocabulary.

Figure 8.2 shows two pages of a translated manual: one page is the English version and the other is the Finnish version. Notice these features:

- Vocabulary is standardized (choke, valve, stop, start).
- Chart layout makes reading easier.
- Symbol, symbol name, and symbol meaning are clearly laid out and reproduced identically from one language to the other.
- English version is on page 3; Finnish version is also on page 3 of the Finnish section.

Figure 8.2. Translated Manual and Standardized Vocabulary and Symbol Explanation (English and Finnish). The chart layout and the one-word system for symbol names plus symbol explanation make translation easier. (Reprinted from *Evinrude® Outboards, 1983* (Waukegan, IL: Outboard Marine Corporation). With permission.)

Symbol	Symbol Name	Meaning or Purpose of Symbol
"Functional Description" Symbols		
	CHOKE	Identifies CHOKE control.
	VALVE	Identifies a control valve.
	STOP	Identifies STOP SWITCH control. May also identify STOP position of throttle control on certain motors.
	START	Identifies position of throttle control device during starting. May also identify STARTING control.
	START-MOTOR	Operating device for starting motor.
"Instructional" Symbols		
	LATCH	Identifies device provided to LATCH or UNLATCH engine cover.
	FUEL SHUT OFF	Identifies device provided to cut off fuel supply to engine.
	SPARK ADVANCE	Number (in degrees) following this symbol indicates recommended maximum spark advance for engine. (Symbol and number appear on engine surface.)
	KEROSENE	Indicates KEROSENE is to be used or identifies KEROSENE is present.
	FUEL	Indicates GASOLINE is to be used or identifies GASOLINE is present.
	OIL	Indicates OIL is to be used or identifies OIL is present.
	FUEL OIL MIX	Identifies FUEL/OIL mixture for 2-stroke engine. Indicates each 50 parts of gasoline are to be mixed with 1 part of oil. Mixture to be mixed completely.

Figure 8.2. (continued)

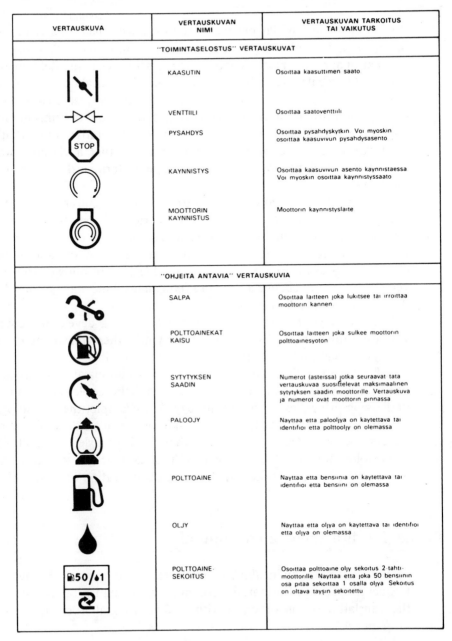

VERTAUSKUVA	VERTAUSKUVAN NIMI	VERTAUSKUVAN TARKOITUS TAI VAIKUTUS
"TOIMINTASELOSTUS" VERTAUSKUVAT		
	KAASUTIN	Osoittaa kaasuttimen saato
	VENTTIILI	Osoittaa saatoventtiili
	PYSAHDYS	Osoittaa pysahdyskytkin Voi myoskin osoittaa kaasuvivun pysahdysasento
	KAYNNISTYS	Osoittaa kaasuvivun asento kaynnistaessa Voi myoskin osoittaa kaynnistyssaato
	MOOTTORIN KAYNNISTUS	Moottorin kaynnistyslaite
"OHJEITA ANTAVIA" VERTAUSKUVIA		
	SALPA	Osoittaa laitteen joka lukitsee tai irroittaa moottorin kannen
	POLTTOAINEKAT KAISU	Osoittaa laitteen joka sulkee moottorin polttoainesyoton
	SYTYTYKSEN SAADIN	Numerot (asteissa) jotka seuraavat tata vertauskuvaa suositielevat maksimaalinen sytytyksen saadin moottorille Vertauskuva ja numerot ovat moottorin pinnassa
	PALOOJY	Nayttaa etta palooljya on kaytettava tai identifioi etta polttooljy on olemassa
	POLTTOAINE	Nayttaa etta bensiinia on kaytettava tai identifioi etta bensiini on olemassa
	OLJY	Nayttaa etta oljya on kaytettava tai identifioi etta oljya on olemassa
	POLTTOAINE-SEKOITUS	Osoittaa polttoaine oljy sekoitus 2-tahti-moottorille Nayttaa etta joka 50 bensiinin osa pitaa sekoittaa 1 osalla oljya Sekoitus on oltava taysin sekoitettu

Cultural Differences

When new or unfamiliar products are introduced, particularly in developing countries, cultural patterns may significantly affect their use. Ask international representatives about their experiences in the field and they will quickly regale you with anecdotes. You will hear about eggs fried on electric irons, refrigerators used as air conditioners (just leave the door open), wood fires built in the cavities of gas ovens, and tire rims mounted backward. One frustrated farm equipment representative found that the only way he could convince illiterate peasants that a chopper was a dangerous machine was to toss a hapless farmyard cat into the mechanism.

Such anecdotes reveal difficulties to be found in cultures where technologies are unfamiliar. Developing countries will present many such instances. In addition, the social structure of these countries may be much more rigidly hierarchical, with clear divisions of labor among classes or castes of people. Thus, one worker may be allowed to drive a vehicle, but not to change a tire, and fixing a machine may be considered demeaning or socially taboo.

In brief, if your product is being marketed in a developing country whose culture is markedly different from yours, then your manuals must be crystal clear and as simple and graphic as possible.

A final word: Many cultures in developing nations are quickly adaptive to new technologies and products, once their people understand the underlying principles of the technology. For example, the technological adaptiveness and dexterity of the Eskimos are legendary. Once they understood that fuel is to an engine what fish is to a sled dog and that a snowmobile was not "dead" when the first tank of gas was gone, they were soon zipping over the ice, skilled in the operation of snowmobiles.

Packaging the Translated Manual

Modular format is an invaluable tool in translated manuals. Manuals are frequently packaged in parallel columns, with English on one side and the translation on the other, or with a "shared visual" format, as shown in Figure 8.3.

When a manual is relatively short and simple, you can make one manual serve for three or four languages by experimenting with foldouts and shared visuals. For example, you can devote the left page to a photo or drawing of the product and the right page to a foldout with French, German, and/or Spanish texts laid out identically and keyed identically to the shared visual. Figure 8.4 shows how a short manual for translation

Figure 8.3. Sample Layout of Multilingual Manual, Showing "Shared" Format. The two sample pages in this figure show typical multilingual layouts. The prose text is often laid out in parallel columns (e.g., English and French) which allow use of single photos or visuals that are shared by the various columns of text.

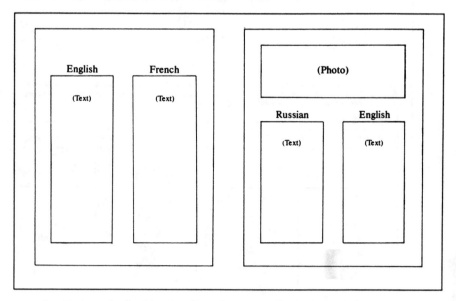

can be formatted to ease translation and keep costs down. The English version of the manual is 18 pages long. The international manual has the following features:

- One international manual, bound as a single book, serves for eight languages (English, German, Italian, Spanish, Dutch, French, Norwegian, Swedish, Danish, Finnish).
- Each language has a separate section (i.e., a French section, a German section, etc.).
- Page layout, number of pages, and page numbering system are identical for each section.
- Shared visuals can be opened out to use with any of the language sections because legends and callout numbers are identical.

On the following pages, a number of examples of creative packaging for international manuals are shown (see Figures 8.4 to 8.8).

Figure 8.4. Translated Manual with Shared, Foldout Visual and Standardized Format and Vocabulary. Note the following features: layout for Italian and English versions are identical (as are the other six language sections included in this manual); page numbering systems

(A) **Incorrect**
OVERLOAD FORWARD
CAUSES BOAT
TO "PLOW"

(B) **Incorrect**
OVERLOAD AFT
CAUSES BOAT
TO "SQUAT"

(C) **Correct**
BALANCED LOAD
GIVES MAXIMUM
PERFORMANCE

⚠ **Safety Warning: If engine is tilted forward so as to cause plowing (see A), swamping may occur in rough water. If engine is tilted aft so as to cause porpoising (see B), steering may be erratic or unstable. See correct angle adjustment (see C).**

Lubrication

TYPES OF LUBRICANT		Contact your DEALER for OMC Lubricants.	
OMC TRIPLE-GUARD™ GREASE		OMC HI-VIS® GEARCASE LUBE	OIL (S.A.E. 30)
TUBE	GREASE GUN		
Ⓐ	Ⓑ	Ⓒ	Ⓓ
LUBRICATION PICTURE SYMBOLS			

LUBRICATION POINTS [30] [31] [32] [33] [34] [35]

30. GEARCASE LUBRICATION
Remove oil drain/fill and oil level plugs from side of gearcase. With motor in normal running position, allow oil to drain completely.
To refill, place tube of OMC HI-VIS® Gearcase Lube or equivalent in drain/fill hole. If OMC HI-VIS Gearcase Lube is not available, OMC Premium Blend Gearcase Lubeor equivalent can be used as an alternate. With motor in normal running position, fill until lubricant appears at oil level hole. See **Specifications** for gearcase capacity.
Install oil level plug before removing lubricant tube from oil drain/fill hole. Drain/fill plug can then be securely installed without oil loss.
If the proper tube or filler type can is not available, install drain/fill plug. Slowly fill gearcase through oil level hole allowing trapped air to escape. Install plug.

A. Oil Level Plug
B. Oil Drain/Fill Plug

Change after first 20 hours of operation and check after 50 hours of operation.
Add lubricant if necessary.
Drain and refill every 100 hours of operation or once each season whichever occurs first.

| Note | Note: Recommended lubricants which have been formulated to protect against damage to bearings and gears must be used as extensive damage can result from improper lubrication. |

31. Idle Speed Adjusting Knob Shaft, Spark Advance Linkage, Cam Roller, Shaft and Gears
32. Swivel Bracket, Engine Cover Latch Shaft
33. Shift Lever Shaft and Detent, Choke and Carburetor Linkage
34. Clamp Screws, Tilt/Run Lever Shaft, Tilt Shaft, Steering Handle, Throttle Shaft and Gears
35. Steering Handle Throttle Gear and Bushing

Frequency of Lubrication	
TYPE OF USE	FREQUENCY
Fresh water	Every 60 days
Salt water	Every 30 days
Storage of 30 days or longer	Before placing in storage

Figure 8.4. (continued). are identical (makes cross-referencing from one language to another easier); the numbers in boxes (30, 31, 32, 33, 34, 35) are lubrication points and correspond to numbers 30–35 on the foldout, shared visual. (Reprinted from *Evinrude® Outboards, 1983* (Waukegan, IL: Outboard Marine Corporation). With permission.)

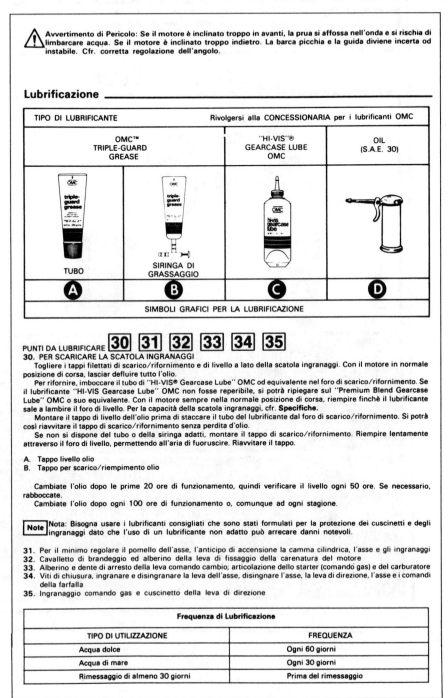

Avvertimento di Pericolo: Se il motore è inclinato troppo in avanti, la prua si affossa nell'onda e si rischia di imbarcare acqua. Se il motore è inclinato troppo indietro. La barca picchia e la guida diviene incerta od instabile. Cfr. corretta regolazione dell'angolo.

Lubrificazione

TIPO DI LUBRIFICANTE		Rivolgersi alla CONCESSIONARIA per i lubrificanti OMC	
OMC™ TRIPLE-GUARD GREASE		"HI-VIS"® GEARCASE LUBE OMC	OIL (S.A.E. 30)
TUBO	SIRINGA DI GRASSAGGIO		
Ⓐ	Ⓑ	Ⓒ	Ⓓ
SIMBOLI GRAFICI PER LA LUBRIFICAZIONE			

PUNTI DA LUBRIFICARE ⟦30⟧ ⟦31⟧ ⟦32⟧ ⟦33⟧ ⟦34⟧ ⟦35⟧

30. PER SCARICARE LA SCATOLA INGRANAGGI

Togliere i tappi filettati di scarico/rifornimento e di livello a lato della scatola ingranaggi. Con il motore in normale posizione di corsa, lasciar defluire tutto l'olio.

Per rifornire, imboccare il tubo di "HI-VIS® Gearcase Lube" OMC od equivalente nel foro di scarico/rifornimento. Se il lubrificante "HI-VIS Gearcase Lube" OMC non fosse reperibile, si potrà ripiegare sul "Premium Blend Gearcase Lube" OMC o suo equivalente. Con il motore sempre nella normale posizione di corsa, riempire finchè il lubrificante sale a lambire il foro di livello. Per la capacità della scatola ingranaggi, cfr. **Specifiche.**

Montare il tappo di livello dell'olio prima di staccare il tubo del lubrificante dal foro di scarico/rifornimento. Si potrà così riavvitare il tappo di scarico/rifornimento senza perdita d'olio.

Se non si dispone del tubo o della siringa adatti, montare il tappo di scarico/rifornimento. Riempire lentamente attraverso il foro di livello, permettendo all'aria di fuoruscire. Riavvitare il tappo.

A. Tappo livello olio
B. Tappo per scarico/riempimento olio

Cambiate l'olio dopo le prime 20 ore di funzionamento, quindi verificare il livello ogni 50 ore. Se necessario, rabboccate.

Cambiate l'olio dopo ogni 100 ore di funzionamento o, comunque ad ogni stagione.

| Note | **Nota:** Bisogna usare i lubrificanti consigliati che sono stati formulati per la protezione dei cuscinetti e degli ingranaggi dato che l'uso di un lubrificante non adatto può arrecare danni notevoli.

31. Per il minimo regolare il pomello dell'asse, l'anticipo di accensione la camma cilindrica, l'asse e gli ingranaggi
32. Cavalletto di brandeggio ed alberino della leva di fissaggio della carenatura del motore
33. Alberino e dente di arresto della leva comando cambio; articolazione dello starter (comando gas) e del carburatore
34. Viti di chiusura, ingranare e disingranare la leva dell'asse, disingnare l'asse, la leva di direzione, l'asse e i comandi della farfalla
35. Ingranaggio comando gas e cuscinetto della leva di direzione

Frequenza di Lubrificazione	
TIPO DI UTILIZZAZIONE	**FREQUENZA**
Acqua dolce	Ogni 60 giorni
Acqua di mare	Ogni 30 giorni
Rimessaggio di almeno 30 giorni	Prima del rimessaggio

Figure 8.5. Bilingual Warning Label. Label uses orange for level of hazard (Warning), a pictogram, and bilingual presentation of all prose text. (Courtesy of Holly A. Webster, Martin Engineering Company, Neponset, IL, 1990. With permission.)

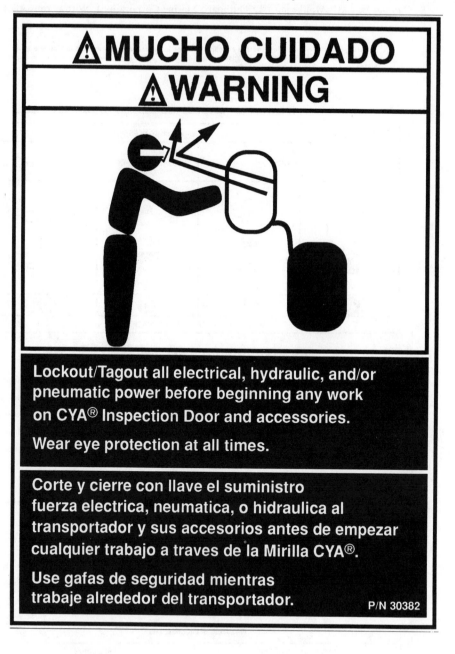

Figure 8.6. Translated Manual Using Color and Modular Format. Colors are used consistently throughout the manual which serves for eight languages. Modular format requires that text be kept short and simple. (Reprinted from *TS 60686* (Allentown, PA: Mack Trucks, Inc., 1986), pp. 20 and 38. With permission.)

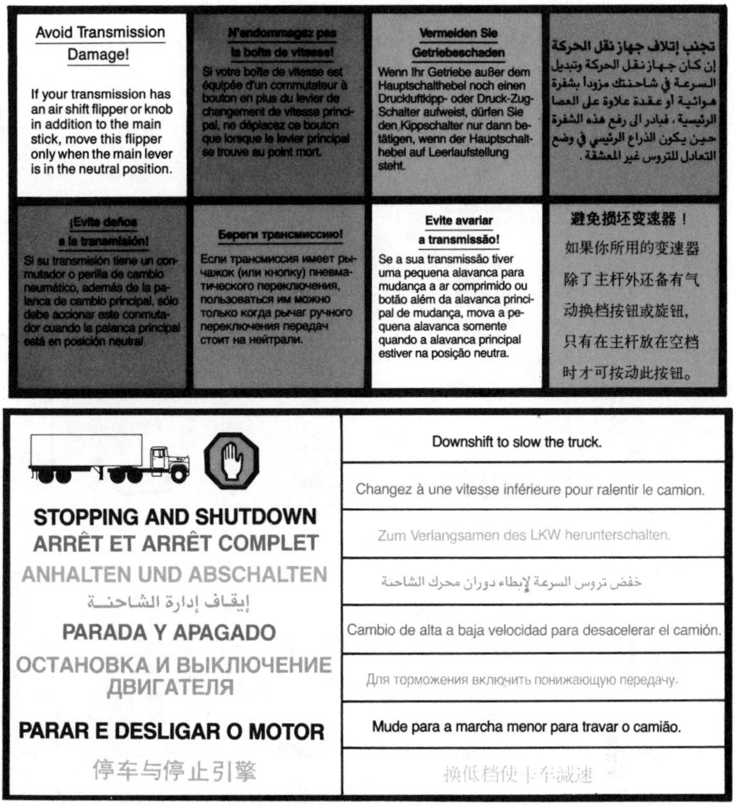

The original pages of Figure 8.6 are in vivid colors. Each language has a designated and unvarying color.

Figure 8.7. International Symbols for Operator Controls. These symbols are standard for automotive and agricultural equipment sold internationally. (Reprinted from *TD8–44/S4 Grain Auger, PU5003/1800* (Ford New Holland, Inc.) With permission.)

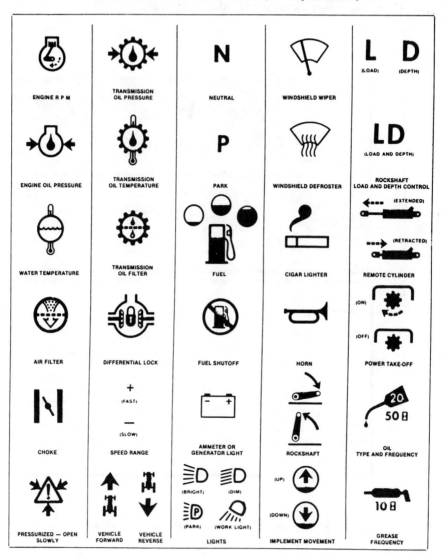

Figure 8.8. Explanation of Standard International Symbols. These symbols can be used on a wide variety of mechanical products. (Reprinted from *Skid Loader 4510/4610 Operators Manual* (West Bend, WI: Gehl Co.), p. 28. With permission.)

STARTER	IGNITION ON	ALL POWER OFF	POWER ON ACCESSORIES	WORK LIGHT	EMERGENCY OR HAZARD FLASHER
HORN	ENGINE OIL PRESSURE	ENGINE HOURS (Running Time)	AMMETER OR GENERATOR LIGHT	HEATER STARTER	FUSE
ENGINE COOLANT TEMPERATURE	Full / 1/2 Full / Empty — VOLUME LEVEL	Gasoline / Diesel — FUEL	CHOKE / ENGINE OIL	HYDRAULIC RESERVOIR	Engaged / Disengaged — HAND BRAKE
CAUTION: PRESSURIZED RADIATOR	NEUTRAL	ROTATE CLOCKWISE	ROTATE COUNTERCLOCKWISE	DIRECTION OF CONTROL LEVER	Float
Fast	REVERSE	FORWARD	FUEL SHUT-OFF	Dump	Lower
Slow — SPEED RANGE	COOLANT	GREASE	DIPSTICK	Roll Back — BUCKET CONTROL	Raise — LIFT ARM CONTROL

Figure 8.9. Special-Order Translations. Manuals in English are sometimes coupled with special-order translations for product accessories such as templates for keyboards, and so on. (Reprinted from *Model M1117A Multichannel Thermal Array Recorder* (Waltham, MA: Hewlett-Packard, 1989), pp. 1–7 and 1–8. With permission.)

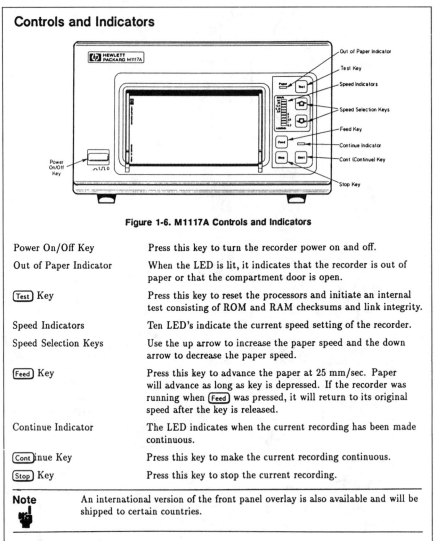

Controls and Indicators

Figure 1-6. M1117A Controls and Indicators

Power On/Off Key	Press this key to turn the recorder power on and off.
Out of Paper Indicator	When the LED is lit, it indicates that the recorder is out of paper or that the compartment door is open.
(Test) Key	Press this key to reset the processors and initiate an internal test consisting of ROM and RAM checksums and link integrity.
Speed Indicators	Ten LED's indicate the current speed setting of the recorder.
Speed Selection Keys	Use the up arrow to increase the paper speed and the down arrow to decrease the paper speed.
(Feed) Key	Press this key to advance the paper at 25 mm/sec. Paper will advance as long as key is depressed. If the recorder was running when (Feed) was pressed, it will return to its original speed after the key is released.
Continue Indicator	The LED indicates when the current recording has been made continuous.
(Cont)inue Key	Press this key to make the current recording continuous.
(Stop) Key	Press this key to stop the current recording.

Note   An international version of the front panel overlay is also available and will be shipped to certain countries.

International Versions

The following items vary according to the country in which the order originated. Appropriate items are supplied for each country.

■ Front Panel Overlays (English or International)

■ Power Cords

■ Fuses

Note   Line voltage switches are set at the factory according to the requirements of the country in which the order originated.

Summary

Whenever manuals must be translated, remember these guidelines:

1. Standardize your vocabulary.
2. Show rather than tell about product and procedures.
3. Use modular format.
4. Package creatively.

9

Managing and Supervising Manual Production

Overview

Manual production cannot be treated in isolation from the company structures that surround it. This chapter is written to help those who are setting up a manual production operation for the first time and those who realize that their present setups do not seem to be working as well as they should.

The quality of service publications depends on a number of key factors: the initial choice and training of writers, the structure and managerial philosophy of the company, the clear delineation of lines of authority, and the fulfillment of writers' basic needs. We include this chapter because we know that the techniques and suggestions we have made to manual writers cannot be effectively applied unless the fundamental company structures are well designed. Good management exists to make it possible for people to do their jobs well.

Who Writes the Manual?

Before we begin a seminar on manual writing, we analyze our participants by asking them to fill out a personal information sheet listing their experience and training for manual writing. Here are just a few of the answers:

> Engineering graduate (all kinds)
> Service and parts manual writer

English or journalism graduate
Law, business, or psychology graduate
Technician
Prototype builder
Company owner
Son of company owner
Transfer from marketing or advertising
Magazine science writer
No experience

Clearly, the entry into specialized technical writing is sometimes through the "front door," but far more often it is an outgrowth of other job duties, a discovery or tapping of a special talent in mid-career, a tangential assignment, or a deliberate second career choice. In small and intermediate-size companies, manual writing is often an add-on to many other job tasks, and writers may be given little guidance on how to proceed.

Choosing the Technical Writer

When a company decides to assign the manual-writing task or to hire a new writer, it often asks, "Should we choose technicians and engineers and then teach them how to write, or should we choose professional writers and teach them the technology of the product?" Posing the question this way can be misleading because of the underlying assumptions — that technicians and engineers can't (or won't) write and that trained writers will probably be technically naive or ignorant.

A better way to think about the choice is to choose someone who *can* communicate and *likes* to. Certainly, some lawyers and engineers are so bound up in jargon that they find it almost impossible to simplify a message for a general public audience, and some wordsmiths write clean and explicit prose, yet are so technically inept that they cannot grasp the workings of the simplest machine. Nevertheless, there are individuals who possess a combination of the necessary communication skills, and these make the best technical writers. They have technical sense about how things work and either know the product from experience elsewhere in the company or can, with a minimum of explanation and hands-on practice, quickly grasp a new technology or product. They recognize clear, correct prose and can also write it. They have a visual sense about drawings, photos, and format devices, and, as we pointed out in Chapter 1, they have "people" skills.

As a manager, you can devise appropriate screening devices to identify good communicators. For example, you can give applicants a simple device or product, along with basic information and relevant photos or drawings. Then ask them to write and lay out a sample page of a manual. Another method is to give applicants sample pages of manuals. Choose a spectrum (some good, some bad, some average) and ask applicants to rank the pages and to justify, in writing, the reasons for their choices.

At the end of the chapter, you will find a reference to a recent listing of academic and technical institutions now offering programs in technical communications. These are helpful if you are looking for new writers.

Writer Training

Writers need training for manual production, just as they do for any other kind of specialized publication. We have found that some companies provide no training whatsoever or give their writers only the sketchiest of orientations, whereas others have systematic and comprehensive training programs.

The best training programs are those provided before manual writing begins, as part of the planning process. After orientation in the fundamentals of manual production, writers are then given periodic training in special skills. In the long run, providing preliminary training is more cost-effective than waiting until writers are floundering. It may then be too late to correct errors, after considerable money has already been spent on such items as art, photography, or printing.

Companies that have been producing products for a long time, especially if they are also large companies, often have sophisticated and comprehensive training programs for manual writers. If you work for such a company, you may already have had orientation sessions, hands-on practice working with other, more experienced writers, and close contact with your publication managers and editors. If your company has no such training program, here are some of the training techniques you might consider:

- Have new writers work through a manual from start to finish with an experienced writer.
- Give new writers in-house handbooks and style guidelines or workbooks to orient them to company procedures.
- Have managing editors work closely with new personnel in the first months on the job.
- When companies are decentralized and manuals are produced at several places, assign one manager to coordinate quality control of

the manuals (Some companies use a "roving editor" who travels among the various manual production locations.)
- Bring writers together for periodic training sessions on such special work topics as metric conversion, the writing of safety warnings, photographic techniques, or desk-top publishing.
- Make libraries and files of company and competitors' manuals available for writers to look at.
- Teach writers "incrementally" by assigning only small segments of a manual for their first assignment, so that they gradually expand their skills and techniques.

If your company is too small or has too few writers to make in-house training sessions cost-effective, you can consider using periodic outside consulting help. Many smaller companies make use of the continuing education conferences and seminars conducted by universities and technical institutes. Working with consultants and attending 2- or 3-day seminars gives writers a chance to learn and exchange ideas on other professionals, to bring themselves up to date on product liability, and to practice their writing skills. These comments, for example, came from writers who had been given outside help, either through consulting or at a conference:

"This is my father's company, and I got the manual-writing job. After attending this conference, I'm going to redo the whole thing. It scares me to see how many mistakes I've made. If someone got injured or killed with our equipment, we wouldn't stand a chance in court with our safety warnings. We're small. We could be wiped out with one lawsuit" (from a writer for small, independent, industrial crane firm).

"I can't believe that I've been doing this job for over three years, and I never knew the range of choices I had for visuals. I didn't know that a manual could be used as legal evidence either. Why didn't somebody tell me?" (from a writer for a large medical equipment company).

"I decided, after this seminar, that we ought to try videotape. I wrote the bible (the script), we hired outside for filming, and my boss was pleasantly surprised. Two tapes that are pretty good for a first try."

If your company makes no provision for organized training sessions, you can still ease the writing process a great deal by creating style handbooks, writer guidelines, and fact sheets listing steps involved in the manual process. Managers or editors can provide such books for their writers, and solo writers can create their own handbooks to systematize the procedures they plan to use. For example, one company's fact sheet, given to writers before work begins on the manual, contains the following information:

- Product name and number
- Deadline dates for completed manuscript in rough form
- Format specifications (column and page width, margins, type size, specifications for photos and drawings)
- Schedule and locations for viewing the product and for hands-on practice with mechanisms of the product
- Style guidelines (average sentence length, vocabulary and language level, use of active-voice verbs, etc.)
- Notifications of what the other segments of the manual will be and which writer is assigned to those segments (especially important for cross-reference work or for machine systems that interact)
- List of phone numbers and names of people who can provide information and of key meetings for product development
- References to materials on file that might be reused

Scheduling and Monitoring Document Preparation

Why Bother with Schedules?

If writing manuals for constantly changing products is difficult, scheduling and tracking these writing projects is even worse. Managers of technical publications departments find themselves often faced with multiple projects, shifting deadlines, tight budgets, and limited staff. In such circumstances it is hardly surprising to find many managers operating in constant crisis mode. Manual writing projects are difficult to schedule for the same reasons that manuals are difficult to write: time and information are usually at least partly out of your control. In addition, the manager has to work within a budget that is dictated from above.

Further, some managers operate under the misconception (sometimes fostered by writers) that writing is a fundamentally different sort of activity from engineering or manufacturing and therefore cannot be planned, scheduled, and monitored using the same methods. Such managers may assign a writer a project and a deadline, but then have no way of assessing progress — unless the writer comes and says he cannot meet the deadline. By then, of course, it is too late to add staff to the project conveniently and the department is back in crisis mode.

In our view, writing manuals is essentially similar to any design activity, with predictable inputs and outputs that can be set up as a series of milestones or laid out in a Gantt chart. Scheduling and monitoring a documentation project not only permits a manager to make adjustments before a project is way behind, but it also permits building a track record that can help in estimating future projects.

How to Develop a Plan

Planning a documentation project requires that you answer three basic questions:

1. What do you need to produce?
2. How much time do you have to work with?
3. What personnel can you use on the project?

These seem obvious, but it is surprising how many managers do not take the time to look systematically for the answers. Let's look at each one.

What Do You Need to Produce? Before you can begin to do any realistic scheduling, you need to have a very clear idea of the nature of the documentation project itself. Are you writing a user's guide or a service manual? Or both? Are you working solely in paper documentation or will you also be responsible for producing or coordinating a videotape? Identify as clearly as possible all the types of documents or the multiple purposes of a single document.

When you know what you are going to produce, begin to plan the documents in more detail. Of course, most of the time this planning activity will be a team effort between manager and writers, but whoever does it, it still needs to be done. This detailed planning stage involves the following kinds of activities:

1. Estimating page count for the final document (including front matter and back matter)
2. Preparing a detailed outline of the document
3. Estimating graphics requirements
4. Identifying tasks required to complete the project (writing, editing, interviewing subject-matter experts, preparing graphics, designing page layout, etc.)

When this stage is finished, you should have a pretty clear idea of the size of the project. It is time to try to fit the project to the time available.

How Much Time Do You Have to Work With? Typically, you will be working under a deadline imposed by someone else, such as the shipping date of the product—determined by marketing. Your job will be to work backward from that deadline to the present to find out how much time is available. Here is the procedure:

1. Backtrack from that shipping date however long it will take for typesetting, printing, binding, and packaging the manual. This is your <u>real</u> deadline.
2. Get out a calendar and count the working days (no weekends or holidays) available between now and then and total them for each month.
3. Multiply the number of days by 6.5 to find the hours available if you put just one person on the job. (Use 6.5 instead of 8 to permit time for meetings, responding to phone calls, etc.)
4. Compare the hours available with your projected page count. From experience you probably have some idea of how many hours it takes to produce a page of final copy (including first draft, second draft, editing and graphics). Typically, companies see a range of values, depending on the complexity of the material and the experience level of staff. Anywhere from 4 to 10 hours per page is pretty common.
5. Use the projected page count and hours available to determine how many staff you will need to assign to the project. For example, if your page count is 150 pages, and it takes 10 hours to produce a page, you will need to have 1500 hours to do the book. If you only have 500 hours available, you will need to assign three staff members to work on the project full time in order to meet the deadline.

What Personnel Can You Assign to the Project? Unfortunately, you will seldom have a full set of writers, editors, graphic artists, and assistants available for assignment. They will all be in various stages of working on other projects. What you have to do, of course, is pull somebody off a project that is nearly finished to get the new project rolling and then add others as they become available.

This kind of juggling act is the "tech pubs" manager's principal activity. It may not be quite as stressful as being an air traffic controller, but it shares the element of needing to keep track of a dozen different things at once. The more systematic you can be about assigning tasks and estimating project requirements, the more smoothly the work will flow and the more likely you will be to meet the deadline. To be systematic, you have to have good information about who is doing what and how far along they are—and that is where project monitoring is essential.

Monitoring an Ongoing Project

Monitoring a project is simply keeping track of how far along the work is and comparing that to the plan, rather like measuring actual expenditures of funds to a budget, and then adjusting for variances. However, how do you know how far along the work is? What are the appropriate milestones? One company uses the following rules of thumb:

- The first draft will take 60% of total time.
- The second draft will take 20% of total time.
- The remaining 20% will go to editing, project management, coordination with graphics, and other support activities.

You know already how many hours you expect the project to take. If you have used up 30% of the time and have only one fourth of the first draft done, you probably need to make some adjustments.

Keeping track of the project requires that the writers and others working on a project log their time and activity. Most companies have some sort of weekly time sheet on which employees show how many hours they spent on which project. Normally these are used to allocate costs among different projects. A time sheet used in conjunction with a brief narrative report of specific activities can give the manager all the information he or she needs to tell whether a project is moving on schedule or not.

Of course, the same information (total project hours, hours per page, etc.) can be used for cost estimating: just multiply hours by the salary (plus overhead) for each employee asssigned to the project.

But Isn't This All a Lot of Record Keeping?

The kind of scheduling and tracking that we discuss here does take some time to set up and keep current. There are computer tools to make it easier, ranging from spreadsheet programs like Lotus 1-2-3 or Excel to project management programs and even at least one software package designed specifically for managing documentation projects (Pubs Estimator). Whatever aids you use, it still does take time.

Is it worth it? The answer to that depends on your situation. If you are a "one-man band," you probably do not need a complex system to track your work. If you are the manager of a 15-person publications department, you probably do need some system to stay on top of progress and recognize the need for adjustments before a crisis develops. If your situation is somewhere in between, you will have to balance the benefits of having the information against the costs of the time it takes to run the system.

The sort of scheduling and monitoring system that we have described does have one additional advantage regardless of the size of your operation: it allows you to build a set of data that you can use to make your estimates more accurate for future projects. If you keep track of the hours per page that a user's manual actually requires, after you have produced three or four of them, you should be able to estimate quite accurately, and you will have hard quantitative evidence to use if you need to lobby for more time or more staff. Your boss may not know about writing manuals, but he or she does know about spreadsheets. You will be speaking a language that is understood in the business world — and that will make your request more credible. Implementing a scheduling system may not solve all your problems of short deadlines and multiple projects, but it will probably make meeting those deadlines a little less chaotic.

Organizational Settings that Affect Writers

The setting and organizational structure in which a writer operates can be the single most important factor in good manual production. Structures affecting manual writers vary enormously from company to company. These variations are sometimes attributable to the size of the company, to managerial philosophy, or to the maturity of the product. Here are some of the patterns that affect the manual writer's job.

Large Companies

The very large company typically has a divisional organization, a diversity of products at scattered geographical locations, and separate cadres of technical writers specializing in manuals for each product category. Further, by the time a company goes national or international, its product line is usually "mature", i.e., the product has been around for some time, and the vocabulary for its parts and systems is quite well established and standardized. (Notable exceptions are the quick-growth electronic and computer technologies and those companies specializing in development of brand-new experimental products.)

Advantages. Very large companies with a team of technical writers whose sole responsibility is manual publication have the luxury of identifying and selecting good communicators from their own ranks, or, when they choose to hire new employees, these companies usually have developed interviewing and testing systems to help them select the most qualified applicants. Quite often, in large companies, writers come up through the ranks, transferring from parts or service manual writing or from positions in product safety, marketing, or advertising. They bring to the job an in-depth knowledge of the product. Large companies are also able, through their service publication managers and editors, to identify writers who need help with their writing skills. That help is provided by one-on-one editorial assistance, on-the-job orientation, and periodic training sessions.

The publication capabilities of large companies often exceed those found in the formal publishing world. Fully equipped photographic labs; sophisticated printing machines; computerized systems for layout, format, and translation and dedicated work stations; full-color duplicating machines; and in-house personnel specializing in art and technical drawing, slide production, film and videotape—all of these are tools of the trade available at many large installations.

Disadvantages. There are also disadvantages to being a manual writer in a large company. Writers may have less autonomy and considerably less flexibility in deciding how best to do their job. If they are at widely scattered locations, they find that information takes longer to travel; filing systems may become harder to tap. If the large company is also decentralized, writing quality may be difficult to control. The manuals produced in Kentucky, for example, may be markedly different in quality and style from those produced in Florida. Because the large company tends to be more rigidly hierarchical, a decision to correct an error or to change the way manuals are done may take years, rather than months, to put into operation. In brief, what is gained through bigness, diversity, and sophistication may be lost through unwieldiness and lack of coordination.

Small and Intermediate-Size Companies

The small and intermediate-size company typically has one or only a few locations. Such companies tend to be regional and centralized and to have a limited product line. Quite often the product is young and innovative and consequently there may be no old manuals to use as guides and no well-established vocabulary for parts and systems.

Advantages. For the writer, the small company can be an exciting and challenging place to work. A young product demands a fresh approach to the manual, and writers can literally be the creator of the vocabulary and the approach. Further, writers are less likely to have to deal with inertia or with "we've always done it this way" frustration. Designers and engineers are likely to be more accessible to answer questions, and decision making is usually more fluid and flexible because the small company hierarchy has fewer layers. In fact, some of the most inventive ideas for manual production and layout come from the small companies lucky enough to have creative writers who had to build a first-time manual from the ground up.

Disadvantages. Many small and intermediate-size companies assign the manual writing to a single individual or to a small group of writers. These writers may be confronted with an awesome array of tasks. They must learn the technology of the product, plan layout, write text and safety messages, arrange for art work, photos, and drawings, negotiate with printers, edit, choose paper stock and typefaces — and often do their own typing and desktop publishing.

Publication support systems may be spotty in the small company (often little more than a typewriter, a desk top, and a corner in a office), and much of the production work must be contracted for. Manual writers who work "solo" feel the pressure of multiple responsibilities and are often rushed and isolated.

Managing the Work Place

Publication managers need to be alert to the importance of the work place, its structure, and its decision-making processes. It is possible to make vast improvements in writers' effectiveness with attention to such details as adequate work space and illumination, systems of information gathering, and acquisition of proper tools of the trade.

Establishing Lines of Authority

Every company has its own internal peculiarities, its hierarchies and pecking orders. Writers work within that pecking order, and situations will inevitably arise in which one person or unit has priority over another. Most writers can live comfortably with lines of authority, if they know what they are. What employees (writers included) find difficult are confused, pass-the-buck procedures in which the lines of authority are never articulated or clearly established.

As a supervisor, you may know where final authority lies for decision on the manual, but you may neglect to convey that information to your writers. You should try to let your writers know about situations in which their decisions are likely to be superseded by someone with higher authority.

In manual writing, the most common problems with lines of authority arise in the following procedures:

- Determining who will have final say and sign-off on the manual's technical accuracy
- Deciding on appropriate language levels for manuals (engineers and lawyers are often disturbed by the simplicity required for general public users)
- Deciding on final authority when distinctions must be made between legal safeguards and engineering safeguards
- Deciding on final authority when an editor and a writer disagree sharply on word choice, format, or stylistic preference

Whenever possible, decisions like these should be made by discussion and consensus, with writers included in the discussion. However, when negotiation is clearly not an option, let writers know where final authority lies.

What the Writer Needs

In Chapter 1, we discussed the writer's two basic needs: information and time. In this chapter, we added a third: training. If you are a manager of service publications, you should provide writers with as much assistance as possible. They should have

- Access to information
- Adequate time to do good work
- Training (either in-house or outside)

We suggest that you reread the suggestions on information gathering and scheduling in Chapter 1. Many of the techniques suggested there involve tasks for which you, as a supervisor, may bear chief responsibility.

Most of the technical writers we talk to are eager to do competent work. They are also quick to sense whether management seems to be "for" or "against" them. As the supervisor of manual production, you should be your writers' chief advocate in insisting on information access, training and time.

Recognition for the Technical Writer

To the list of basic writers' needs, we must add a fourth: *recognition*. Times are changing. For many companies, the manual used to be regarded as a bothersome necessity that got written at the last minute.

Such attitudes were reflected in the scant time and money allotted to the manual and the meager recognition given to writers.

Industries are beginning to recognize that manual writers are the bridge builders between the product and the consumer. After your product has left the dealer's store, the manual becomes the interpreter of your product. Without the manual, the consumer must make a phone call or a trip back to the seller for help. As products grow more complex, as formerly simple mechanical devices are steadily being electronically controlled and computerized, and as manuals are interpreted by courts as significant legal evidence, the technical writer's work is becoming more valuable — and more highly valued. The new attitudes are reflected in better salaries, more investment in writing training programs, and better integration of the technical writer into the mainstream of company organizational structures.

The coalition of technical writers into a cohesive profession has also begun to take shape with conferences, seminars, professional societies, newsletters, and books. The Society for Technical Communication is a national organization with regional and local chapters. Its publications help writers stay abreast of the field and keep in touch with each other. We hope that this book will prove to be a help to writers who have chosen technical writing as their profession.

For Further Reading

1. Society for Technical Communication, 815 Fifteenth Street N.W., Washington, D.C. 20005, (202) 737–0035.
2. *INTERCOM*, October 1989 (list of technical communication programs).

Index

DATE		
APR. 03 1995		
07		